Environmental Science, Engineering and Technology



# Abiotic Stress

## New Research

# ENVIRONMENTAL SCIENCE, ENGINEERING AND TECHNOLOGY

Additional books in this series can be found on Nova's website under the Series tab.

Additional E-books in this series can be found on Nova's website under the E-book tab.

# ENVIRONMENTAL REMEDIATION TECHNOLOGIES, REGULATIONS AND SAFETY

Additional books in this series can be found on Nova's website under the Series tab.

Additional E-books in this series can be found on Nova's website under the E-book tab.

ENVIRONMENTAL SCIENCE, ENGINEERING AND TECHNOLOGY

# ABIOTIC STRESS

## NEW RESEARCH

NIKHIL HARYANA
AND
SHREYA PUNJ
EDITORS

Nova Science Publishers, Inc.
*New York*

For permission to use material from this book please contact us:
Telephone 631-231-7269; Fax 631-231-8175
Web Site: http://www.novapublishers.com

## NOTICE TO THE READER



## LIBRARY OF CONGRESS CATALOGING-IN-PUBLICATION DATA

Abiotic stress : new research / editors, Nikhil Haryana and Shreya Punj.
p. cm.
Includes index.
ISBN 978-1-61942-197-4 (hardcover)
1. Plants--Effect of stress on--Research. 2. Plants--Adaptation--Research. 3. Plant physiology--Research. 4. Plant genomes--Research. I. Haryana, Nikhil. II. Punj, Shreya.
QK754.A265 2011
571.2--dc23
2011043384

*Published by Nova Science Publishers, Inc. † New York*

# CONTENTS

# PREFACE

Abiotic stress is the negative impact of non-living factors on the living organisms of a specific environment. The non-living variable must influence the environment beyond its normal range of variation to adversely affect the population performance or individual physiology of the organism in a significant way. Abiotic stresses, such as salinity, drought, temperature, and oxidative stress, cause major loss in crop yields. Topics discussed in this book include drought and nutritional stresses in plants; NADP-dependent enzymes and abiotic stress; circadian regulation of abiotic stress tolerance in caenorhabditis elegans; and plant cortical microtubules and abiotic stress signals.

Chapter 1- Diverse environmental stresses are connected to soil-derived limitations (e.g. drought, salinity, and nutrient deficiencies). They are often interconnected, exposing crops to multiple stresses and inducing similar damage to plants. The manner in which plant roots sense and respond to different environmental factors appears to be overlapping. These stresses are inducing a series of morphological, physiological, biochemical and molecular changes in the root system, striving to adapt to the stress conditions and/or to modify the environment from which the stress is originated. This adjustment includes the ability to exude a wide array of micro- and macromolecules into the rhizosphere having the potential to affect and alter the relationships between plants and soil microorganisms. Indeed, through the exudation of a wide variety of compounds, roots may regulate the soil microbial community in their immediate vicinity, encourage beneficial symbioses, change the chemical and physical properties of the soil, and inhibit the growth of competing species.

Chapter 2- Abiotic stress often negatively affects photosynthesis, which is the main source of NADPH. Reduced form of the coenzyme represents an indispensable compound for biosynthetic reactions, antioxidant systems, and for enzymes involved in regulation. As the ratio $NADP^+$/NADPH in plants influences redox reactions, the control of cell redox homeostasis is important for balancing metabolic processes and redox-dependent signalling.

NADPH is produced mainly by enzymes from oxidative pentose phosphate pathway (glucose-6-phosphate dehydrogenase, 6-phosphogluconate dehydrogenase) but also other enzymes (e.g. NADP-isocitrate dehydrogenase, NADP-malic enzyme, NADP-glutamate dehydrogenase, and non-phosphorylating NADP-glyceraldehyde-3-phosphate dehydrogenase) represent an alternative source of NADPH in stressed plants. NADP-glutamate dehydrogenase is involved in catabolism of proteins and amino acids and thus provides NADPH. However, the role of this enzyme could be different under stress condition (e.g. in the case of increased protein synthesis). Reaction catalyzed with non-phosphorylating NADP-glyceraldehyde-3-phosphate dehydrogenase drives NADPH/$NADP^+$ ratio to biosynthetic processes.

Chapter 3- As sessile organisms plants have to cope with changing environmental conditions. Drought and salinity, which causes water-deficit in plant cells, are common adverse factors that limit plant growth and productivity. Understanding the mechanisms by which plants perceive environmental signals and transmit them to cellular machinery to activate adaptive responses is of great importance to biology and to rational engineering of crop plants. This chapter reviews the signal transduction mechanisms that activate water-deficit stress responses and the regulation of transcription factors that control the expression of stress-responsive genes. The general components of stress signal transduction pathway for drought and salt stress are considered. Signal perception, receptor-coupled phosphorelay, phosphoinositol-induced $Ca^{2+}$ changes, $Ca^{2+}$-coupled phosphoprotein cascades, mitogen-activated protein kinase cascade and transcriptional activation of stress responsive genes are the main signal transduction steps addressed. Abscisic acid (ABA) plays a pivotal role in stress responses in plants. Therefore, the hormone implications are also reviewed. The transcription factors responsible for reprogramming gene expression in response to stress are described. Finally, the physiological and biochemical responses that lead to plant tolerance to water-deficit stress are addressed.

Chapter 4- In all ecosystems, plants continuously face environmental stress and consequently are forced to respond with defensive and adaptive strategies. These responses require the activation of several signaling pathways that induce expression of specific genes. An effective response requires that the biological system have the genetic background to support the necessary molecular players that permit the assembly of essential integrative genetic pathways. In the most complicated scenario, plants must contend against more than one abiotic stress, pest, or pathogen at the same time, forcing an integral and complete defense response to adjust plant physiology. Certain molecular players act as hubs or master regulators to integrate signals from different regulatory pathways activated by two or more types of abiotic stress or forms of biotic stress. Some defense and adaptive genes are transcriptionally regulated not only by a specific abiotic stress but also by biotic stress, exposing the existence of overlapping pathways. The genes shared between these networks appear to allow plants to prioritize their responses and ensure their survival by using their resources efficiently. The well-documented shared or convergence points in the response to abiotic and biotic stress in plants exhibit a superimposed complexity, as exemplified by induction of certain defense genes by biotic stress in presence only of a specific environmental condition (temperature and humidity). Interestingly, accumulating data strongly support the hypothesis that the convergence points between abiotic and biotic stress pathways also modulate the post-embryonic developmental program which is one of the most conspicuous adaptive strategies to cope with environmental stress. Pathogen attack or abiotic stresses such as nutrient scarcity alter cell division and cell differentiation processes, and consequently the plant architecture is modified. Plant growth regulators, such as auxins, cytokinins, ethylene, and jamonic acid, as well as reactive oxygen species, play crucial roles in the early steps of the convergence between these multiple stress signals. The advancement in powerful molecular tools, including transcriptome and proteome analysis, whole-genome sequencing, and bioinformatic studies are enabling disection of networks in abiotic and biotic signaling cascades and identification of the overlapping reactions and key factors that fulfill very important roles as integrative signals in plants.

Chapter 5- Circadian rhythms are ubiquitously found in nature. These rhythms are generated and sustained by a conserved time-keeping system that coordinates the biochemistry, physiology and behavior of an organism with daily changes in the environment. There is increasing evidence that this coordination contributes to the overall health of organisms. Hence, it is of great importance to understand how the circadian clock is able to transduce environmental signals to internal processes. One of the outputs of the circadian system in invertebrates, vertebrates, bacteria and plants is the stress resistance behavior. In this work the authors focus on the circadian variation in tolerance to several abiotic stressors in the nematode *Caenorhabditis elegans*. Circadian and daily variations in response to osmotic-, oxidative- and cyanide-mediated stress have been described at the behavioral and molecular level.

Overall, the biological clock helps *C. elegans* cope with naturally occurring rhythms in abiotic stressors present in its natural habitat. It provides tools for preparing the organism to defend itself by predicting when the defense mechanisms are most needed.

Chapter 6- Plant growth and productivity are affected by environmental stresses, including drought, high salinity, and high and low temperature. Plants percept the stress signals and response to adapt these stresses. Plant abiotic stress responses are regulated by the integration of various environmental and endogenous signals including plant hormones. Plant adaptation to abiotic stresses is controlled by molecular cascades, in which expression of many genes are modulated, and also involved in the cross-talk pathway with other signaling mechanism. In this chapter, the authors show an overview of the current knowledge on plant molecular responses and adaptation to abiotic stresses.

Chapter 7- High soil salinity, drought, and low temperature response mechanisms have been studied intensively. Recently, significant progress has been made in elucidating the nature of microtubules (MTs) that are believed to be intimately involved in stress response mechanisms. In recent years, the known functions of MTs have expanded from being just static scaffolds to playing critical roles in almost all intracellular activities. Accordingly, research on MT arrays in the context of abiotic stresses may reveal important mechanisms of plant stress tolerance, and consequently these arrays have become a hot and key research topic. In this chapter, the authors review current knowledge on the relationship between plant cortical microtubules (CMTs) and abiotic stress signalling. The data show that at least three early response mechanisms, namely, abscisic acid (ABA) biosynthesis, calcium ion (Ca2+) signalling, and the formation of plant cell walls, follow CMT reorganization, and are mediated by dynamic changes in the CMTs.

In: Abiotic Stress: New Research
Editors: Nikhil Haryana and Shreya Punj
ISBN: 978-1- 61942-197-4


*Chapter 1*

# DROUGHT AND NUTRITIONAL STRESSES IN PLANT: ALLEVIATING ROLE OF RHIZOSPHERIC MICROORGANISMS

***Laura Bardi*[*a] *and Eligio Malusà*[a,b]**
[a]CRA-RPS Agricultural Research Council, Research Centre for the Soil-Plant System - Turin Research Group, Environment Park, Turin, Italy
[b]Research Institute of Horticulture, Skierniewice, Poland

## ABSTRACT

Diverse environmental stresses are connected to soil-derived limitations (e.g. drought, salinity, and nutrient deficiencies). They are often interconnected, exposing crops to multiple stresses and inducing similar damage to plants. The manner in which plant roots sense and respond to different environmental factors appears to be overlapping. These stresses are inducing a series of morphological, physiological, biochemical and molecular changes in the root system, striving to adapt to the stress conditions and/or to modify the environment from which the stress is originated. This adjustment includes the ability to exude a wide array of micro- and macromolecules into the rhizosphere having the potential to affect and alter the relationships between plants and soil microorganisms. Indeed,

* CRA-RPS Agricultural Research Council, Research Centre for the Soil-Plant System - Turin Research Group, Environment Park - Via Livorno, 60 (A2 laboratori), 10144 Turin, Italy, laura.bardi@entecra.it, http://rps.entecra.it , Tel. +39-011-2257281/7280, Fax. +39-011-2257281

through the exudation of a wide variety of compounds, roots may regulate the soil microbial community in their immediate vicinity, encourage beneficial symbioses, change the chemical and physical properties of the soil, and inhibit the growth of competing species.

Plant-soil feedback processes are an important aspect of the plant responses to environmental stresses. The evidence of feedback is strongest for plants growing in extreme environments or under stressing conditions, and in case of plant-soil microorganisms interactions. For example, establishment of symbioses with arbuscular mycorrhizas, which can alter water movement in soils by promoting hydraulic lift and enhancing plant water uptake, may be utilized by plants to cope with water scarcity. Numerous species of plant growth–promoting bacteria have been observed to form biofilm when colonizing roots. The advantages of bacterial biofilm formation for the plant derives from the improvement in bio-geochemical processes leading to increased availability of sparingly soluble mineral nutrients as well as to the production of nitrogen compounds.

The presence of phosphate transporters in arbuscular mycorrhizal fungi, that take up inorganic phosphate from the soil, together with phosphate transporters present in plants that are mycorrhiza-specific, confirm the intimate interrelation between the root and symbiotic fungi. However, the increased evidence that bacteria-fungi interactions are more widespread than expected, points out that their dynamics may be crucial for plants under environmental conditions inducing abiotic stresses to plants. The picture is complicated by the effects on both rhizobacteria and mycorrhizas of protozoan grazers, which are also directly affecting the root system.

Understanding the interactions between microorganisms living in the rhizosphere and the plant root system is essential for understanding how abiotic stresses affect the plant-soil-microrganisms relationships. A practical outcome from this knowledge would allow the inoculation of plants with not autochthonous, specifically selected beneficial rhizospheric microorganisms as a possible tool to increase plant tolerance to common abiotic stresses.

Recent advances in elucidating the role of the interactions between plant roots and microorganisms present in the rhizosphere under the constraints derived by limited water availability and nutritional deficiencies are reported.

## 1. INTRODUCTION

The inability of plants to move forces them to adapt to a range of abiotic and biotic stresses. Variation in soil nutrient composition is a common condition in natural environments, but also in agricultural systems. Indeed, only 30–50% of applied N fertilizers and 10–45% of P fertilizers are taken up by crops (Adesemoye and Kloepper 2009). Furthermore, there is a general trend worldwide in reducing chemical inputs in agriculture due to environmental concerns. Reduction of fertilizers inputs or use of organic and less soluble fertilizers long afterwards, as well as a limited availability of nutrients in low fertility soils and of fertilizers in several developing countries, could represent a possible source for nutrient stresses on agricultural crops.

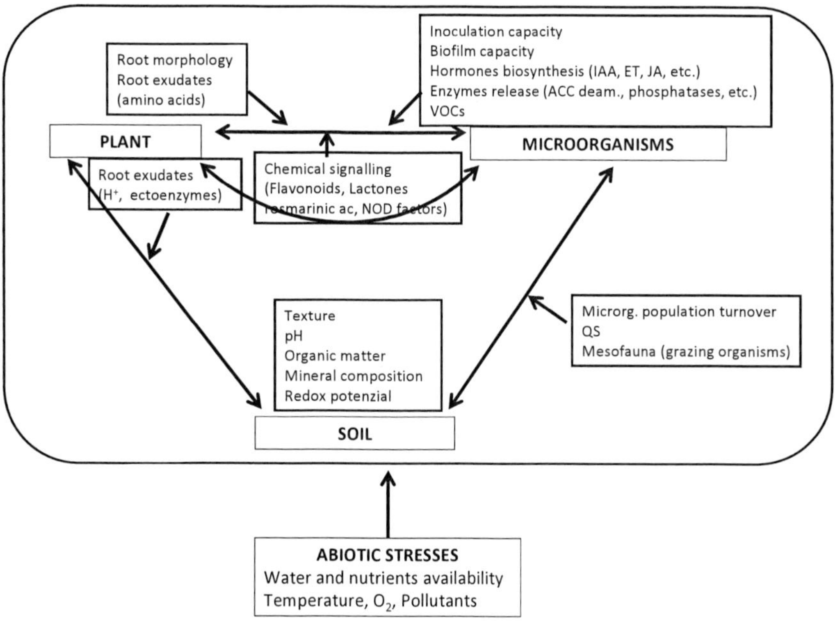


Figure 1. The interactions between plants, microorganisms and soil are affected by the genetic and metabolic characteristics of the two groups of living organisms and by the physic-chemical properties of the soil. The interactions between plants and microorganisms, as well as among different microorganisms, are based on a chemical signaling that regulates them, frequently on the basis of a feedback system. This system affects also the mechanisms utilized by plants and microorganisms in relation to soil characteristics.

Water availability for agriculture is a permanent problem in hot-arid countries. Water deficiency is an environmental and a social problem, as it severely limits crops yields and causes competition between agricultural and other human uses. Many studies are devoted to improve drought tolerance of plants in order to reduce water needs and to increase yield in arid areas of the world. Nowadays this is also a rising problem in temperate countries, prompting more attention to any intervention capable to increase water use efficiency in plants by biotechnological or agronomical ways.

The acquisition of water and nutrients from soils is governed by roots and their interaction with the abiotic and biotic components of soil. Rhizosphere is the interface where the soil, plant roots, microbes and fauna strongly interact (Hartmann et al. 2008). This “space” is further classified into ectorhizosphere, the portion ouside the root, and endorhizosphere, which includes the root epidermis and cortex, where endophytic microorganisms can be present. The surface of the root system and any strongly adhering soil particle are defined as rhizoplane (Kennedy 1998). Figure 1 summarizes the different interactions established within the rhizosphere and the main factors that are affecting them.

The soil biotic component include complex communities with different trophic levels that function in very heterogeneous environments (Giri et al. 2005) formed by several kinds of microorganisms that are living symbiotically or in association with roots. Among the symbionts, two groups of microorganisms are of paramount importance for plant growth: arbuscular mycorrhizal fungi (AMF) and N-fixing nodulating bacteria. Non-symbiotic plant growth promoting microorganisms are classified as rhizospheric or endophytic depending on their presence in the rhizosphere or inside the plant tissues, respectively. The bacteria living freely in the rhizosphere or in association with plants and that are promoting the growth of plants by different mechanisms are termed plant growth promoting rhizobacteria (PGPR) (Vessey 2003). Population densities of bacteria in the rhizosphere may be up to 1000-fold higher than in bulk soil and up to 15% of the root surface may be covered by micro-colonies of a variety of bacterial strains.

All these microorganisms might enhance acquisition of nutrients in low-nutrient soils and alleviate the effects of low water availability. However, a full appraisal of their role in relieving the stress posed to the plant by these conditions has begun to emerge only recently. The term ‘induced systemic tolerance’ (IST) for PGPR-induced changes in plants resulting in enhanced tolerance to abiotic stress was recently proposed by Yang et al. (2009); differences in functionality of AMF as a result of abiotic stresses have prompted the need to reconsider some well known data into a more integrated framework (Smith et al 2010).

This chapter is reviewing information on the support provided by rhizosphere soil microorganisms to plants in coping with stresses derived from nutrient and water availability. For exhausting lists of microbial strains that have been selected and studied for improving plant nutrition and/or drougth tolerance the reader can refer to few recent reviews (Augé 2001; Ruiz-Lozano 2003; Vessey 2003; Saleem et al. 2007; Hayat et al. 2010; Kang et al. 2010; Yang et al. 2010, Smith et al. 2010)

## 2. Plant-Microorganisms Interactions and Effects on Nutritional Stresses

Under nutrients limiting conditions, growth is reduced and plants modify their physiology and morphology to cope with scarce resources. These modifications depend on the specific nutrient mobility in soil (Hodge 2004). The first response to nutrient deficiency appears to be more related to general stresses than specifically to nutrient deprivation (Hammond et al. 2003). Indeed, plants have the ability to remobilize major nutrients from the pool present in the cell (potassium from the vacuole, phosphorus and nitrogen from molecules containing these minerals or the vacuole). However, commonly to most abiotic stresses, the production of reactive oxygen species (ROS) in roots is a general feature in response to nutrients deprivation (Shin et al 2005; Shin, Schachtman 2004; Malusà et al. 2002; Juszczuk et al. 2001). The generation of ROS, such as superoxide (O2•−), hydrogen peroxide ($H_2O_2$) and hydroxyl radical (•OH) can cause rapid and deleterious oxidation of most cell components such as proteins and membrane lipids or may cause DNA injury (Apel and Hirt 2004). Therefore, plants have evolved a dynamic antioxidant defence network. The enzymes super oxide dismutase, ascorbate peroxidase, catalase, together with ascorbic acid and glutathione (GSH) are considered the major ROS scavenging mechanisms in plants (Mittler 2002). The regulation and interaction between ROS, GSH and ascorbate impact the synthesis of plant hormones such as gibberellins, abscisic acid, and ethylene, which may signal plant response to nutrient deficiency (Schachtman and Shin 2007). Oxidative stress regulation is considered to play an important role among the different mechanisms proposed to explain the ability of AMF to alleviate the effects of different abiotic stresses affecting plant growth and yield production. (Ruiz-Lozano 2003) Strong induction of antioxidative enzymes during the colonization period resulted in detoxification of ROS (generated during colonization) in mycorrhizal shrubs under drought stress (Alguacil et al. 2003). Water stressed AM colonized roots of citrus plants showed higher activities of superoxide dismutase, catalase and ascorbate peroxidase as well as ascorbate when compared with non-AM roots (Wu et al. 2006).

Changes in root morphology or architecture in response to nutrient starvation involve the hormones auxin and ethylene (Lopez-Bucio et al. 2002; Nacry et al. 2005), compounds that might be produced or which synthesis might be affected by rhizosphere and endophytic microorganisms. Roots exposed to abiotic stimuli often respond by producing more ethylene, a typical stress-related hormone, than under normal conditions, which results in retardation of their growth (Morgan and

Drew 1997). The direct precursor of ethylene in the plant biosynthetic pathway is a cyclic amino acid, 1-aminocyclopropane-1-carboxylic acid (ACC), that can be hydrolyzed by several endophytic and rhizobacteria that possess the enzyme ACC deaminase (Blaha et al. 2006). ACC deaminase degrades ACC into α-ketobutyrate and ammonia (Honma, Shimomura 1978) which can be used by the bacteria as a source of nitrogen and carbon for growth (Klee et al. 1991). In this way, the bacteria acting as a sink for ACC are inducing a reduction of ethylene level in the plant and can prevent some of the potentially deleterious consequences of high ethylene concentrations (Glick et al. 1998; Saleem et al. 2007).

A model of the action of the enzyme to interpret its wide presence in rhizosphere bacteria, in the light of increased plant tolerance to abiotic stresses, has been proposed by Glick et al. (2007). The model takes into account the interactions between auxin and ethylene in both PGPR and plants. Indeed, indole-3-acetic acid (IAA) is the most widespread hormone produced among rhizobacteria (Patten and Glick 1996), and endophytic PGPR (Tan and Zou 2001). Using tryptophan excreted by plant roots to the rhizosphere, PGPR synthesize IAA which can be taken up by roots. IAA promotes root growth, but also enhance the activity of ACC synthase. The increased concentration of ACC molecules in root cells leads to the release of ACC into the soil which becomes available to rhizosphere microbial communities expressing ACC deaminase. Growth of such strains is accelerated in the close vicinities of plant roots in comparison to other soil microorganisms (Glick et al. 1998). The release and metabolization of ACC reduce the biosynthesis of the stress hormone ethylene in the roots, thus reducing its negative influence on plant growth. Interestingly, it has been found that the growth promotion effect of microbial-derived ACC deaminase activity is most effective in stress environments (Cheng et al. 2007). This has been supported by data showing that PGPR with ACC deaminase trait used as inoculants give very consistent results in improving plant growth and yield (Burd et al. 2000; Shaharoona et al. 2006)

A possible side effect of the presence of PGPR expressing ACC deaminase is in enhancing AM colonization. The negative effect of ethylene on plant colonization by AMF was firstly reported in the symbiotic association between *Medicago sativa* and *Glomus mosseae* (Azcon-Aguilar et al. 1981), and subsequently confirmed with other fungal and plant species (*G. ramisporohora* colonizing *Poncirus trifolata* and *G. aggregatum* colonizing *Pisum sativum*) (Ishii et al. 1996; Geil and Guinel 2002). On the other hand, the effect of AMF roots inoculation on ethylene synthesis has been contradictory. Ethylene formation was increased in roots of *Linum usitatissimum* inoculated with *G. intraradices*

(Dugassa et al. 1996), but not in tomato inoculated with *G. mosseae* (Vierheilig et al. 1994).

Ethylene is an inhibitor of rhizobial nodulation of legumes. However, many rhizobia strains produce a low level of ACC deaminase which allows them lowering the ethylene levels and increase nodulation by 25–40% (Ma et al. 2003, 2004), thus prompting the speculation that the enzyme activity facilitates nodulation by locally lowering ethylene levels (Glick et al. 2007). ACC deaminase activity of rhizobacteria promoted nodulation in several legumes (Ma et al. 2003; Dey et al. 2004; Okazaki et al. 2004).

Colonization of various plants with mycorrhizal fungi leads to increased endogenous levels of jasmonates, plant hormonal regulators involved in plant responses to abiotic stresses (Wasternack and Hause 2002), within the roots (Vierheilig and Piche´ 2002). Jasmonates are also involved in the signal transduction pathway activated upon interaction of plants with microorganisms (Pozo et al. 2004). Several roles have been suggested for jasmonates in these interactions (Hause et al. 2007):

a) induction of flavonoid biosynthesis to act as signals to stimulate the growth of AM fungi;
b) reorganization of the cytoskeleton in arbuscule-containing and non-colonized cortical root cells;
c) alterations of the sink status of the roots;
d) increase in plant fitness mediated by effects on the level of cytokinins.

With respect to this latter aspect, it is worth noting that about 90% of microorganisms found in the rhizosphere are capable of releasing over 30 growth promoting compounds of the cytokinin group when cultured in vitro (Barea et al. 1976).

Biosynthesis and accumulation of phenolic compounds is another response of plants to different abiotic stresses including nutrient starvation (Dixon and Paiva 1995). Phosphorus, sulphur, iron, calcium and magnesium starvations stimulate the production of phenolics in the plant tissues (Cesco et al. 2010) and a marked increase in the activity of several enzymes of phenylpropanoid metabolism (Sánchez et al. 2000; Malusà et al. 2006). These compounds may be released in the root exudates and act as signal molecules that govern the interactions between plants and soil microorganisms involved in nutrition (Werner 2001; Bonfante and Requena 2011). Furthermore, plants can produce and secrete various compounds that mimic quorum sensing signals of bacteria, a major regulatory mechanism in bacteria to adjust their metabolism to crowded conditions (Whitehead et al. 2001), therefore altering bacterial activities in the rhizosphere (Bauer and Mathesius 2004).

## 3. Role of Soil Microorganisms in Reducing Nutritional Stresses

**Box 1. Mechanisms of plant nutrients uptake affected by soil microorganisms**

i. increase in the surface area of roots by a direct extension of existing root systems (e.g. through mycorrhizal associations)
ii. enhancement of root growth, branching and/or root hair development (e.g. through hormones produced by plant growth promoting rhizobacteria),
iii. direct contribution to nutrient availability through
    a. N fixation (e.g. rhizobia and non-symbiotic diazotrophs)
    b. stimulation and/or contribution to metabolic processes that affect mobilization of nutrients from poorly available sources (e.g. release/exudation of organic anions)
iv. indirect effect on nutrient availability by
    a. displacement of sorption equilibrium that results in increased net transfer of nutrients into soil solution
    b. transformation of nutrients among different pools (microbial-mediated processes that alter the distribution of nutrients within and between the different inorganic and organic forms, for example nitrification/denitrification)
    c. the turnover of microbial biomass within rhizosphere

Microorganisms are able to stimulate plant nutrient uptake through several mechanisms that are acting directly or indirectly on the plants (Box 1: Richardson et al. 2009). Other mechanisms that are affecting plant growth, and indirectly may reverberate on the ability of the plant in nutrient uptake, include enhanced resistance to drought (Alvarez et al. 1996), salinity, waterlogging (Saleem et al. 2007), the ability to produce ACC deaminase to reduce the level of ethylene in the root of developing plants thereby increasing the root length and growth (Li et al. 2000; Penrose and Glick 2001), the ability to produce hormones like auxines

(IAA) (Patten and Glick 2002), abscisic acid (ABA) (Dobbelaere et al. 2003), gibberellic acid (GA) and cytokinins (Dey et al. 2004), antagonism against phytophatogenic bacteria by producing enzymes, antibiotics, fluorescent pigment and cyanide (Jetiyanon and Kloepper 2002; Zehnder et al. 2001), enhanced resistance to oxidative stress (Stajner et al. 1995, 1997), production of water-soluble B group vitamins (Revillas et al. 2000).

Following are reviewed the roles of different microorganisms interacting with plants and reducing the stress derived by the major growth limiting nutrients.

### 3.1. Nitrogen

The role of AMF in nitrogen uptake has been relatively recently appreciated, in comparison to phosphorus nutrition (Govindarajulu et al. 2005; Marschner and Dell 1994). Even though the high rate of N uptake by mass flow could indicate that AM fungi may not be that important in the uptake of inorganic N by the host plant (Liu et al. 2007; Atul-Nayyar et al. 2009), more than 50% of plant N requirement may be supplied by mycorrhizal association (Govindarajulu et al. 2005; Leigh et al. 2009). Such ability of AM fungi is particularly important for plants growing under conditions with limitations, for example when specific agricultural methods of production are applied (e.g. organic farming where chemical fertilizers are not allowed) or under arid and semi arid conditions, where water availability limits uptake also of mobile nutrients such as inorganic N (Subramanian and Charest 1999; Tobar et al. 1994a; Hodge and Fitter 2010). Indeed, due to the narrower diameter of the hyphae of AM fungi (in the range 2–20 μm) in comparison to that of root hairs (Drew et al. 2003), AM plants are able to access nutrients from soils having small pores not accessible to roots, which are typical of dry or compacted soils. Increased activity of enzymes involved in N metabolism (i.e. nitrate reductase and glutamine synthase) in both roots and shoots of mycorrhizal plants can also indirectly enhance plant resistance to drought stress or alleviate its effects (Subramanian and Charest 1999).

AM fungi take up both types of inorganic N sources present in soil solution, e.g. ammonium (Johansen et al. 1993) and nitrate (Hawkins et al. 2000). The colonization by AM fungi activates plant ammonium transporters suggesting a similarity between N and P transport in mycorrhizal plants (Guether et al. 2009). However, differences in this respect between plant species have been pointed out. One or two AM-inducible ammonium transporter genes of the AMT family have been found in *L. japonicus* and *M. truncatula* symbiosis (Benedito et al. 2010), while five AM-inducible transporters occur in soybean mycorrhized roots (Kobae et al 2010), pointing to a possible different spatial distribution or functional divergence of the transporters.

AMF are able to regulate, probably indirectly, plant N uptake under different soil concentrations. A medium level of N fertilization resulted in a higher N uptake from mycorrhizal plants with respect to high or low N fertilization rates (Azcón et al. 2001, 2008). However, the availability of high amounts of inorganic N in soil, particularly ammonia, significantly reduces the rate of mycorrhiza colonization of roots and therefore decreases the contribution of the mycorrhiza to N uptake (Azcon et al. 1992; Valentine et al. 2002).

AM fungi can transfer substantial amounts of nitrogen forms to the host plants, utilizing the inorganic N released from organic sources (Hamel 2004). Organic forms of N are also directly taken up in the form of amino acids (Hawkins et al. 2000) using a specific transporter expressed in the extra radical hyphae (Cappellazzo et al. 2008).

The growth of fungal hyphae in patches with active organic matter decomposition (Aristizabal et al. 2004), may increase the effectiveness of N supply to the host plant (Hodge et al. 2001a, b; Hodge and Fitter 2010) also in conditions without inorganic N fertilization (e.g. organic farming). Nevertheless, the presence of inorganic and organic N sources modifies the response of mycorrhizal plants. Wheat colonized by *G. mosseae* absorbed organic N in the form of N-glycine by 0.2% and 6% at low and high level of N-fertilization, respectively (Hawkins et al. 2000). The higher organic N uptake with a sufficient N fertilization was attributed to the more developed hyphal network. Both organic and inorganic N forms are transferred from the extra radical hypha to the plant in the form of arginine (Tian et al. 2010).

It has been recognized that the concomitant presence of multiple AM fungal species results in a more effective residue decomposition and N uptake than when a single AM fungal species is present (Tu et al. 2006). Indeed, different AM fungal species vary markedly in their root colonization and mycelium growth, leading to variation in their ability to supply plants with nutrients and promote plant growth (van der Heijden et al. 2003). This can be especially true for those plants associated with AM fungi developing a hyphal network relatively close to the roots, such as *Scutellospora calospora* (Smith et al. 2000). *G. fasciculatum* resulted more efficient in nitrate uptake and translocation than *G. mosseae* because the former had a higher percentage of root infection and more extensive hyphae (Azcon et al. 2001).

Diazotrophs (i.e. $N_2$-fixing bacteria) are of utmost importance in supporting N plant nutrition, particularly in conditions of limited availability. They are classified as being either symbiotic (rhizobia and *Frankia* species) or free living or associated to the plant without establishing a symbiosis, such as endophytic bacteria (Sturz et al. 2000; Cocking 2003).

The rhizobium—legume symbiosis fully embodies the concept of plant nutrient stress relief by soil microorganisms: in nitrogen-poor environments the *Rhizobium* promotes legume plant growth by providing a limiting nutrient. Symbiotic rhizobia can provide up to 90% of the N requirements of the host legumes through atmospheric $N_2$ fixation (Franche et al. 2009), by the activity of nitrogenase present in rhizobial bacteroids, which converts atmospheric elemental nitrogen into ammonia (Shiferaw et al. 2004). Rhizobia nodulation is affected by several factors related to the interaction between the host and the microsymbiont, the physico-chemical conditions of the soil, the presence of mycorrhizal fungi and the presence of bio-molecules such as flavonoids, polysaccharides and hormones (Tisdale et al. 1990; Zafar-ul-Hye et al. 2007) (Fig. 1).

Rhizobia can also develop non-specific associative interactions with roots of non-legumes and promote their growth, and as such are also commonly considered as PGPR (Sessitsch et al. 2002). They have also been isolated as natural endophytes from roots of several non-legumes species such as maize, wheat, rice and canola (Hayat et al. 2010; Yanni et al. 2001) either when these species were grown in rotation with legumes or in a mixed cropping system involving symbiotic legumes.

Several non symbiotic N-fixing bacteria species have been proved to enhance N uptake of plants. *Azospirillum* species grow extensively in the rhizosphere of gramineous plants (Kennedy and Tchan 1992; Kennedy et al. 2004). They enhance both development and yield of different host plants, but this improvements are attributed mainly to an effect on root development which results in increased water and mineral uptake, and to a lesser extent to biological $N_2$-fixation (Okon and Labandera-Gonzalez 1994; Okon and Itzigsohn 1995). Nevertheless, plant inoculated with *Azospirillum brasilense* promoted greater uptake of nitrate (and also of potassium and phosphorus) in corn, sorghum and wheat (Zavalin et al. 1998; Saubidet et al. 2000), also under field conditions (Galal et al. 2000; Panwar and Singh 2000). N from biological fixation in wheat plants inoculated with *A. brasilense* and *A. lipoferum* was in the range of 7–12% (Malik et al. 2002), while in sugarcane up to 60–80% N derived from N-fixation mainly by *A. diazotrophicus* (Boddey et al.1991).

Among other non-symbiotic diazotrophs that have been found increasing N nutrition can be mentioned *Burkholderia vietnamiensis* and *Herbaspirillum* sp. The former allowed to reduce by 25–30 kgN $ha^{-1}$ the dose of N fertilizer when rice was inoculated under field conditions (Tran Vân et al. 2000). *Herbaspirillum*, an endophyte bacterium, was able to provide rice plants with N from biological fixation in the range 19.5-58.2% of total plant N, depending on plant age and cultivar (Baldani et al. 2000; Mirza et al. 2000). Several *Pseudomonas* species and other bacterial strains isolated from a calcareous soil had a significant positive effect on cotton and pea N content (and also P and K content) in nutrient-poor soil (Egamberdiyeva, Höflich 2004).

Nevertheless, the provision of N to the plant by the free living rhizosphere bacteria has been questioned after finding that N-fixing PGPR strains lacking nitrogenase activity retained the ability to promote plant growth (Hurek et al. 1994; Adesimore et al 2009), thus suggesting the occurrence of other mechanism/s of growth promotion. Among them, the increase of N use efficiency seems to play a major role: PGPR-elicited plant growth promotion was associated to a significantly better N uptake per gram of tissue and N uptake on a whole-plant basis in tomato plants receiving a lower dose of fertilizer in comparison to the recommended one (Adesemoye et al. 2008; Adesimore et al. 2009).

A model integrating the plant and inoculants activities to explain the enhancement of N uptake efficiency in plants has been proposed (Adesimore et al. 2009). PGPR promote the growth of the plant increasing the root surface area or modifying the general root architecture (Lucy et al. 2004). Better growing plants in turn release higher amounts of C in root exudates which prompts increase in microbial activity, and this process continues in a cycle. Such cycle of events make more N available for plants' uptake. The model did not infer which mechanisms could be at the basis of the activity of PGPR. However, the modification of the root system could be induced by either the release of hormones (e.g. auxines) or the reduction of ethylene production by metabolizing ACC produced by the plant.

The efficiency of different strains on different crops in N uptake can vary. N content in wheat plants inoculated with *A. brasilense* decreased as N supply rate increased, and at the maximum N supply, the content of total N was the same between inoculated and non-inoculated wheat plants (Saubidet 2002). On the other hand, Shaharoona et al. (2008) reported that N use efficiency increased in response to inoculation with *Pseudomonas fluorescens* at all fertilizer levels in wheat. The differences in effectiveness of different bacterial inoculants to induce such effects could be linked to the differences in biofilm formation and in survival rate of each strain (Jayasinghearachchi, Seneviratne 2006).

Besides the direct provision of N through biological fixation or uptake, bacteria and AMF may increase indirectly the availability of organic N forms, by hydrolysis of complex organic molecules, or of inorganic N, by mineralization. Breakdown of proteins and peptides from soil organic matter and microbial biomass turnover as well as exudation by roots make up the pool of amino acids in the soil solution which account for about 50% of total soluble N (ranging in concentrations from 0.1 to 50 mM) (Jones et al. 2002). Nevertheless, even though the significance of soluble organic N for plant nutrition has been demonstrated for soils in a range of different ecosystems (Schimel and Bennett 2004), the slow movement of amino acids in soil relative to nitrate, the rapid turnover of amino acids by soil microorganisms, and the poor competitive ability of plant roots to capture amino acids from the soil solution result in limited availability of organic N in agricultural soils. Plant roots could only utilize on average 6% of the added amino acid with the rest captured by the microbial biomass (Owen and Jones 2001).

Mineralization of organic forms of N to ammonium and its subsequent nitrification to nitrate is a microbial-mediated process that affects both N availability and rhizosphere dynamics. The process is carried out by different groups of bacteria, but interactions with the plant and other microorganisms is affecting the rate of the process through a 'priming effect' provided by both microorganisms and plant (Kuzyakov 2002). Nevertheless, this effect could be affected by the organic matter composition, particularly the C:N ratio, and the compounds forming the organic C pool as well as the plant species (Tuomela et al. 2000). It has been shown that the AM fungus *Glomus hoi* can enhance the mineralization process of the soil organic matter leading to an increase of available N (Hodge et al. 2001a). However, several factors can influence the process of N mineralization in the presence of AMF:

a) the extension of extraradical hyphae, which is related to fungal genome, soil conditions and plant-fungus interactions;
b) the capacity of production of enzymes degrading the soil organic matter (e.g. pectinase, and cellulase)
c) the interactions between AMF and other soil microbes (e.g. protozoa and nematodes) and soil mesofauna (e.g. earthworms).

On the other hand, the influence of AMF hyphae on the activity of soil mineralizing microbes can also indirectly affect soil N dynamic and uptake by plant (Marschner et al. 2001; Hodge 2003a, b). Interactions between AM fungus and other microorganisms leading to changes in N mineralization can be directly induced by the volume of the hyphae since AM hyphae are an important source of organic C for soil microbes (Schimel and Weintraub 2003), or indirectly by the modification induced by AMF of the plant root exudates which are then affecting the microbial populations hosted in the mycorrhizosphere (Toljander et al. 2007). Stimulation of AM fungal growth by N inputs enhanced decomposition of newly added residues in soil (Tu et al. 2006), which could be promoted by stimulating microbial activity through enhanced labile C availability from the host plant in the mycorrhizosphere (Zak et al. 2000). Support to this effect can be seen from the significant decrease in the C/N ratio of the soil induced by the presence of AM fungi (Atul-Nayyar et al. 2009). The presence of bacterial and fungal-feeding organisms has been shown to enhance N flows in the rhizosphere (Bonkowski et al. 2000). Indeed, about 60% of protozoa and nematodes feeding material is excreted and made available for microbial mineralization (Bonkowski 2004). However, the interaction could also lead to changes in the composition and activity of the microbial community. Bacterial and fungal feeding nematodes populations are increased in soils rich in organic matter or fertilized with organic fertilizers and consortia of AMF and PGPR (Malusà et al. 2011).

### 3.2. Phosphorus

Phosphorus is a major limiting nutrient for plant productivity, mainly because of its high reactivity with iron, aluminium and calcium in soils, which can result in precipitation of up to 90% of the soil phosphorus (Gyaneshwar et al. 2002). It derives that normally in soil the concentration of soluble P forms taken up by plants (orthophosphate anions, predominantly as $HPO_4^{2-}$ and $H_2PO_4^{1-}$) is usually very low: typically 1 to 5 μM (Bieleski 1973). However, plants are able to uptake P from solutions of concentration also ranging from 0.01 to 0.1 μM (Jungk 2001).

Plants have several strategies to adapt to a low-P environment:

a) changes in root architecture resulting in rapid rate of root elongation and high root to shoot biomass ratio (Hill et al. 2006), increased root branching and root angle (Lynch and Brown 2001), formation of root hairs (Gahoonia and Nielsen 2003);
b) mobilization of sparingly soluble inorganic and organic P through modification of the rhizosphere environment by acidification, exudation of organic acids and phosphatases (Ryan et al. 2001; Duff et al. 1994);
c) modification of expression of high-affinity phosphate transporter genes (Smith 2002);
d) enhancement of P uptake capability through establishment of mycorrhiza symbiosis (Read and Smith 2008);
e) exploitation of PGPR capable of solubilising inorganic or organic P (Hartmann et al. 2009).

Even though only the last two strategies seem related to plant-microorganisms interactions, also the first three are affected by or can shape the relationship of plant with both AMF and PGPR, influencing the plant response to the P-starvation stress. The increased root development under P starvation conditions results in a higher root surface available for the establishment of rhizosphere microorganisms or provide more sites for mycorrhiza inoculation (Gamalero et al. 2004).

Amount and composition of the exudated organic anions (also driven by root-microorganisms interactions – Hartmann et al. 2009) and their ratio of metabolization by rhizosphere microorganisms are all factors affecting the efficacy of the organic anions in nutrient mobilization. On the other hand, these exudates may increase the population of rhizosphere bacteria that are able to mobilize poorly soluble nutrients (Jones et al. 2003). Exudates from fungal hyphae solubilised more P than root exudates alone, suggesting that the mycorrhiza contributed to increased P uptake through solubilisation of inorganic P forms (Tawaraya et al. 2006).

A nonspecific mycorrhizal-responsive acid phosphatase was found in the *Tagetes patula*–AM fungal association (Ezawa and Yoshida 1994). The gene *TpPAP1,* encoding the mycorrhiza-responsive acid phosphatase, with increased transcript levels in response to AM formation, was cloned and it was found that it is regulated at the gene expression level (Ezawa et al. 2005). On the basis of such findings it was proposed that the fungal partner activates host phosphatase secretion as a part of the low-P adaptation system of the plant partner.

AM hyphae also excrete phosphatases able to hydrolyze organic P, but this mechanism is considered of minor importance for the uptake of P (Joner et al. 2000; Richardson et al 2009). Furthermore, an increase in extractable organic P was found to be associated to increase in microbial biomass around the roots and considered a possible consequence of microbial-mediated immobilization of orthophosphate within the rhizosphere (Richardson et al. 2005).

The higher uptake of P by mycorrhizal plants is commonly explained in terms of increased hyphal exploitation of the soil (Schneph et al. 2008) and their ability to absorb it from localized sources (Cavagnaro et al. 2005). The presence of the Pi depletion zone in the rhizosphere is a major factor contributing to the advantage of plants forming mycorrhizal associations. While in non-mycorrhizal roots the extension of the Pi depletion zone is closely related to root hair length, in mycorrhizal roots the depletion zone of Pi greatly exceeds the root hair cylinder (Smith and Read 2008).

It has been established that mycorrhizal plants have two pathways for P uptake: the direct pathway through root hairs, and the indirect pathway through fungal hyphae (Smith et al. 2004). Poulsen et al. (2005) provided evidence that the two pathways operate simultaneously and expression of the respective Pi transporter genes is involved. However, AM formation reduces expression level of the Pi-transporter genes of the host plant (Glassop et al. 2005; Paszkowski et al. 2002), while the fungal Pi transporter on extraradical hyphae plays a pivotal role in P uptake (Liu et al. 1998) suggesting that in mycorrhizal plants P uptake proceeds mainly through the AM pathway (Burleigh et al. 2002). Interestingly, mycorrhizal plants showed lower Km values for orthophosphate uptake than non-mycorrhizal roots (Cardoso et al. 2006). However, a functional diversity in mycorrhizal phosphate uptake, expressed by the effect that the specific plant and fungal species association have on the contribution of both the plant and the mycorrhizal Pi uptake pathways to the total P uptake, has been observed (Smith et al. 2003, 2004).

The AMF and plant P uptake pathways have been shown to function in a not additive manner (Smith et al. 2010). Indeed, even though several field and glasshouse experiments have shown that the establishment of AM symbioses in low-P soil produces an increase in plant P uptake and growth in comparison to non mycorrhizal plants (Smith and Read 2008), in some cases the response of the plant to AM colonization has not resulted in an increase in total plant P (Smith et al. 2003, 2004). Nevertheless, P uptake through the AM pathway to plants occurs also in these cases as it has been confirmed showing that expression of genes for individual P transporters in root cortical cells are up-regulated in AM plants, regardless of the plant responsiveness to the AMF colonization (Bucher 2006; Poulsen et al. 2005). These findings have prompted a new broader vision of the ecological and physiological role of mycorrhizal symbiosis, as a strategy to respond to abiotic stresses, such as N starvation (see above) or drought (see below), or to increase the fitness of the plant modifying the soil structure or its tolerance to pathogens (Smith et al. 2010). In this respect, it is worthy to note that plant responsiveness to AMF inoculation depends on several factors deriving from the genome of the two symbionts, their interaction and, probably, from the genome and population structure of the bacteria interfacing both plant and fungus (Johansson et al. 2004).

Phosphate-solubilising microorganisms include a wide range of non-symbiotic organisms that are able to solubilise poorly soluble forms of phosphorus (Gyaneshwar et al. 2002; Whitelaw 2000).

*Pseudomonas* and *Bacillus* strains are very effective in solubilise P from precipitated inorganic sources such as tricalcium phosphate, hydroxyl apatite and rock phosphate (Rodríguez and Fraga 1999; Banerjee et al. 2006). The mechanism of solubilisation relies on the production of several different organic acids (Rodríguez and Fraga 1999; Hayat et al. 2010). The efficiency in P solubilisation depends on the C and N source available in soil and also on the buffer capacity of the soil (Illmer and Schinner 1995; Whitelaw et al. 1999).

Several species of rhizobia are able to solubilise both organic (Abd-Alla 1994; Chabot et al. 1996a) and inorganic phosphates (Antoun et al. 1998) even though sometimes the efficacy of P solubilisation is lower than that of *Pseudomonas* and *Bacillus* species (Alikhani et al. 2006).

Direct P mobilization has been shown to be performed also by non-symbiotic fungi such as *Penicillium* sp. (Wakelin et al. 2007). Fungi are more effective at solubilisation of Fe and Al phosphates (Whitelaw 2000).

Several bacteria strains are able to release P from organic sources (Rodríguez and Fraga 1999). Microbial mediated mineralization of organic P occurs by secretion of different enzymes: phosphatase, phytase, phosphonoacetate hydrolase, D-α-glycerophosphatase and C-P lyase (Hayat et al. 2010; Rodríguez and Fraga 1999). Such activity contributes significantly to plant availability, since organic forms of P constitute around 50% of the total P in soil and their greater mobility in solution (Seeling and Zasoski 1993) makes them the predominant form of P found in soil solutions (Ron Vaz et al. 1993).

The microbial biomass and microbial turnover are also key aspects for the supply of P (Seeling and Zososki 1993). The bacteria populations turnover is affected by the activity of other microorganisms such as nematodes and protozoa (Bonkowski 2004), which might then also influence the overall P availability to plants.

Even though improved P uptake of plants inoculated with P-solubilizing microorganisms is attributed to the P-solubilising activity of the microorganisms involved, and P-solubilising bacteria strains of *Bacillus* and *Paenibacillus* have been applied to soils to specifically enhance the phosphorus status of plants (Brown 1974), their effect under field conditions has been debated considering, as the main factor for the inconsistent results, the insufficient knowledge on the dynamics of microorganisms and their interactions (Richardson 2001).

## 3.3. Other Nutrient Elements

Uptake from soil of mineral elements by the mycorrhizal mycelium and subsequent transfer to the plant have been demonstrated also for essential micronutrients such as Zn and Cu (Guo et al. 1996), and, depending on soil pH, also for K, Ca and Mg (Clark and Zeto 2000; George et al. 1992). However, these elements are generally present in soil in sufficient concentration, therefore the contribution of soil microorganisms to their uptake might be considered marginal.

Nevertheless, *Trichoderma harzianum* Rifai 1295-22, a fungus known for its biocontrol activity (Howell 2003), was able to solubilise sparingly soluble minerals ($MnO_2$, metallic zinc, and calcium phosphate) by chelation and reduction (Altomare et al. 1999). The role of micronutrients in root stress tolerance and plant disease resistance is less well known, but it is gaining increasing recognition since many micronutrients are required in the pathway of polyphenols formation and are present in the molecules of ROS-scavenging enzymes (Scandalios 1993), hence they are crucial for plant stress tolerance capabilities.

Iron is, among the micronutrients, frequently causing deficiencies symptoms and stress due to its low solubility or content in soil, particularly in alkaline and calcareous soils (Masalha et al. 2000). AM fungi seem to increase also Fe uptake, but this depends on several factors: plant species and genome (Clark and Zeto 1996b), soil pH (Clark and Zeto 1996a), soil supply of P and soil temperature (Raju et al. 1990). Furthermore, the hyphal uptake of iron by mycorrhizal plants appears to be regulated by the Fe nutritional status of the plants: with long-term low Fe availability, plants obtained more Fe via their mycorrhizal hyphae (Caris et al. 1998). The influence of all these factors could explain the contrasting behaviour that has been recorded in case of iron uptake with different species of mycorrhizal plants: soybean was showing either a decrease (Pacovsky and Fuller 1988) or an increase (Clark and Zeto 1996b; Lambert et al. 1979) in shoots Fe concentration after mycorrhization. Maize and sorghum showed an increases in Fe concentration in shoots after inoculation with mycorrhiza (Lambert et al. 1979; Caris et al. 1998), while peanuts did not (Caris et al. 1998).

Among PGPR, fluorescence *Pseudomonas* species synthesize a group of proteins, pyoverdines, which are a class of siderophores with high affinity for Fe(III) (Meyer and Abdallah 1978). Pyoverdines are only expressed under iron deficient conditions due to the high cost of their synthesis by the bacterial cell machinery (Visca et al. 2007). They are released into the soil and after forming the Fe-pyoverdine complex are transported into the cell by receptors present on the outer membrane (Faraldo-Gomez and Sansom 2003).

The effect of PGPR on plant Fe uptake depends on the kind of strategy the plant is using to assure its uptake. Dicotyledonous and non-graminaceous monocotyledonous plants employ a iron acquisition mechanism based on the reduction of Fe in the rhizosphere through enhanced proton extrusion (named Strategy I). In contrast, graminaceous plants release phytosiderophores that solubilise $Fe^{3+}$ and chelate it into complexes that are taken up by the root cells via a specific plasma membrane transport system (Strategy II) (Marschner and Römheld 1994). The latter mechanism is similar to the microbial siderophore strategy of Fe uptake (Guerinot 1994). For Strategy I plants, modification of the iron bioavailability in the rhizophere by the plant can lead to the selection of specific bacterial populations which are adapted to iron depletion through the production of specific siderophores. These populations can affect directly plant growth by promoting its iron nutrition since it has been shown that Strategy I plants species can absorb microbial $Fe^{3+}$-siderophores complexes probably by endocytosis (Vansuyt et al. 2007). However, no beneficial effect of inoculation with siderophore-producing (pyoverdine) *P. putida* in comparison with a siderophore-negative mutant of *P. putida* and a siderophore-auxotroph *Arthrobacter flavescens* strain was observed in maize and cucumber plants (Walter et al. 1994). Nevertheless, Fe acquisition may be indirectly affected, by increasing the extracellular supply of Fe at the root surface (Walter et al. 1994) or by decreasing the growth of phytopathogen saprophytic through pyoverdine-mediated iron competition (Lemanceau et al. 2009). For strategy II plants, it has been proposed that possible degradation of Fe-pyoverdines by microbes could lead to the release of iron that would then be available for phytosiderophores (Duijff et al. 1994b).

It is worthy to note that also certain species of *Rhizobium* and *Bradyrhizobium* can release siderophore which could be involved in iron or microelements plant nutrition (Plessner et al. 1993; Jadhav et al. 1994).

## 3.4. Interactions between Different Soil Microorganisms and Nutrient Stresses

The interactions between bacteria and AM fungi have potentially beneficial functions related to nutrient deficiency stresses, particularly when PGPR (Kloepper 1996; Barea et al. 2002) and $N_2$-fixing bacteria (Secilia and Bagyaraj 1987; Biró et al. 2000) are involved.

The interaction can be direct or indirect. Direct modification of bacterial population composition and activity by AM fungi have been ascribed to competition for inorganic nutrients (Christensen and Jakobsen 1993). The population of autotrophic nitrifying bacteria in pot cultures of sweet corn colonised by the AM fungi *G. mosseae* and *G. fasciculatum* was significantly higher than in non-mycorrhizal cultures, and the mycorrhization induced a reduction in ammonifying and denitrifying bacterial populations (Amora-Lazcano et al. 1998).

Indirect effects to associated bacteria can be modulated through the changes of the chemical composition of root exudates induced by AMF colonization (Gryndler 2000). Changes in the rhizosphere of mycorrhizal plants can affect microbial populations both quantitatively and qualitatively and the mycorrhizosphere generally has features that differ substantially from those of non-mycorrhizal plants (Barea et al. 2002). A wide range of microorganisms (including actinomycetes and various PGPR) associate with mycorrhizas within the mycorrhizosphere (Rillig et al. 2006; Toljander et al. 2007). In some cases, a degree of specificity in the response of some bacterial species to certain AMF species has been found (Andrade et al. 1997). Besides, a more common association of Gram-positive bacteria with AM fungi than Gram-negative bacteria has been observed (Artursson et al. 2005). Such findings are suggesting that bacteria could be activated by species-specific fungal exudates. Anyway, the different mechanisms proposed to explain such effect relate to the increased rate of nutrients cycling due to enhanced soil microbial content and microorganisms biodiversity found in soils where mycorrhizal plants are grown (Gange 2000).

Legumes can form tripartite symbiotic associations with nodule-inducing rhizobia and AM fungi simultaneously, which increase both P and N uptake efficiency (Lisette et al. 2003). AM fungi are known to enhance nodulation and N fixation by legumes (Barea et al. 2005b). Mycorrhizal and nodule symbioses often act synergistically on infection rate, mineral nutrition and plant growth. In the legume tripartite system, AM establishes the symbiosis with the host roots more rapidly than nodules do (Mortimer et al. 2009), and may therefore be able to influence the effects of external N supply on nodule development and functioning. AM colonization increased the proportion of nitrate reductase allocated to roots in alfalfa co-inoculated with *Sinorhizobium* with a more pronounced effect when nitrogen was absent from the medium (Vazquez et al. 2001).

Both rhizobia and AMF can affect the inoculation and nodulation capacity of the other associated microorganism, respectively. AMF colonization was enhanced by co-inoculation with rhizobia, especially under low P conditions and AMF improved soybean nodulation, especially under low P and low N conditions (Wang et al. 2011). Synergic effects of co-inoculation with rhizobia and AM fungi have been also reported for faba bean (Li et al. 2004). Increased N and P uptake in soybean was reported after co-inoculation with rhizobia and the mycorrhizal fungus *G. mosseae* (Xie et al. 1995). $N_2$ fixation rates of alfalfa plants inoculated with *Rhizobium meliloti* were higher in mycorrhizal plants than in non-mycorrhizal plants (Toro et al. 1998)

The plant-*Rhizobium* system benefits from the presence of AM fungi because the mycorrhiza ameliorates not only P uptake but also increases the availability of other nutrients that might be limiting *Rhizobium* (Pacovsky 1986; Smith 2002). The better nutritional status of the plant would improve photosynthesis efficiency, making a greater proportion of photosynthates available to the rhizobium nodules.

This is supported by the fact that when non-mycorrhizal plants are supplied with a readily available P source, their $N_2$-fixing ability is similar to mycorrhizal plants (Karandashov and Bucher 2005). The uptake of other essential micronutrients from the soil by the AM fungal hyphae might also play an indirect effect upon the $N_2$-fixing system since enzymes for N-fixation contain several microelements (Seefeldt et al. 2004).

Wang et al. (2010) examined the effects of different parameters such as plant genotype and N and P levels on the tripartite symbiosis between soybean, AM fungi and *Bradyrhizobium japonicum*. Co-inoculation with rhizobia and AM fungi significantly increased soybean growth under low P and/or low N conditions, as indicated by increased plant N and P content, than when singly inoculated with rhizobia or AMF. Contrasting results in P uptake were related to root architecture features (i.e. depending on plant genome). Both AMF colonization and nodulation were more responsive to P availability in a deep-rooted genotype in comparison to a P-efficient shallow-rooted genotype (Wang et al. 2011).

Dual inoculation with *Azotobacter chroococcum* and *G. mosseae* resulted in a significantly higher amount of other nutrients (K, Ca, Mg, Cu, Fe, Zn and Mn), and higher biomass production, in comparison to that of un-inoculated or single inoculated pomegranate plants also under dry conditions (Aseri et al. 2008).

The benefit of the presence of both symbionts is evident also when external nitrogen is supplied. Mycorrhizal bean plants co-inoculated with *Rhizobium* induced an increase of N fixed with external ammonium supply and it was related to lowering accumulation of asparagines, which is a feedback inhibitor on biological N-fixation (Mortimer et al. 2011).

Increased $N_2$ fixation and plant N content was also reported in several legumes co-inoculated with *Rhizobium* sp. and PGPR (Parmar and Dadarwal 1999; Sivaramaiah et al. 2007). The effect is explained by the stimulation of PGPR of root growth through IAA production that may provide more infection sites for the establishment of the symbiosis.

Interaction between AM fungi and various PGPR has also been reported and in many cases shown to be beneficial for mineral uptake. Inoculation of maize and ryegrass with *A. brasilense* and AMF resulted in N and P contents comparable to plants grown receiving a fertilizer (Adesimore et al 2009; Barea et al. 2002). However, increased N acquisition by dual-inoculated plants was attributed to greater N uptake capacity by mycorrhizal infected roots, rather than a direct effect through $N_2$ fixation.

The interaction between AMF and N-fixing bacteria can also include the association of the endosymbiont bacteria *Burkholderia* sp. with AM fungi (Bianciotto et al. 2000). Genes for N fixation have been localised within the bacteria genome (Minerdi et al. 2001). This could further enhance the capacity of AMF in supporting plant N nutrition in stress conditions.

The interaction between AMF and rhizosphere microorganisms may also involve free-living phosphate-solubilising bacteria. The latter may support P uptake by increasing the soil phosphate pool available for uptake by the extra radical AM fungal hyphae. On the other hand, the mycorrhizal symbiosis may also increase the size of the phosphate-solubilising bacterial population (Kim et al. 1998; Toro et al. 1997). Interactions between AMF and PGPR might also affect microelements uptake: increased uptake was found when a consortium of AMF and PGPR was applied to strawberry plants (Malusà et al. 2007).

## 4. Drought Stress Effects on Plant

The effects of drought on plants and the mechanisms of resistance have been recently reviewed by Farooq et al. (2009).

Plant responses to drought stresses are partially specific, partially aspecific and common to several other stress causes; moreover, water deficit is frequently associated to other abiotic stress factors, such as soil salinity, heat, or nutrients deficiency.

The main consequences of the exposition to drought stress are reduced germination, growth and yield, and eventually plant death. When a plant is exposed to drought it shows physiological and morphological changes, aimed to improve its tolerance or to avoid the stress (Bray 1997).

As soon as the water availability decreases, the turgor potential of the plant cells diminishes; as a consequence, many turgor-driven processes, such as leaf expansion and root elongation, slow down.

Reduced leaf expansion leads to a smaller leaf area, and consequently a decrease in the overall amount of water transpired by the plant. Until a certain level of water deficiency, such strategy allows the plant to limit water uptake without modifying the leaf transpiration rate: this is important to maintain other important functions of transpiration in leaf, such as the regulation of the leaf temperature, and the maintenance of the driving forces that generate the water flux from which the nutrient uptake from soil results. Another strategy to reduce the leaf area, when the drought stress affects a completely developed plant, is to induce the ethylene-mediated leaf senescence and abscission.

The root growth is also influenced by water deficit. Reduced cell turgor potential in root apices located in dry layers of soil decreases root elongation; as a consequence, roots tend to elongate where the water is more available, and this usually happens in the deeper layers of soil, that are less rapidly subjected to drying. The quality (distribution and structure) of roots is considered more important than their volume to determine the efficiency in extracting water from soil (Farooq et al. 2009). Moreover, even if the leaf area index is reduced under water deficiency, the photosynthetic activity is much less affected; as a consequence, there is a higher amount of carbon assimilated in leafs that becomes available to produce root biomass: so the root/shoot ratio is increased. Such additional strategy of plant to cope with drought stress is more effective during vegetative phases of growth than during reproduction: when assimilates are mainly directed to fruits than to roots the plants are more sensitive to drought.

Also auxins can play a role inducing the formation of new roots, but the endogenous synthesis of auxins is usually inhibited by drought. However, in maize an increase of indole-3-butyric acid was observed under water deficit and indole-3-butyric acid synthetase was reported as drought inducible in *Arabidopsis* (Ludwig-Muller 2007).

The water use efficiency (WUE), that is defined by the ratio between the number of the carbon dioxide molecules assimilated by photosynthesis and the number of molecules of water transpired, is usually improved in plants under drought stress. The increase in WUE allows to maintain the growth while reducing the water loss, and it was attributed to reduced transpiration through stomatal closure (Abbate et al. 2004; DeLucia et al. 1989; Lazaridou et al. 2003): in fact, the effect of stomatal closure is more incisive in reducing transpiration than in decreasing the intercellular concentration of carbon dioxide. However, when drought stress persists, WUE dramatically decreases because photosynthesis is impaired due to mesophyll dehydration. In general, the effect of drought stress on stomatal conductance is considered more significant than the direct effect on photosynthesis.

Stomata closure is the most rapid response of plants to drought to protect themself from desiccation, that prevents photosynthesis and consequently growth and yield. Stomata closure can be a passive event (hydropassive stomatal closure), induced by the loss of water from the guard cells by direct evaporation to the atmosphere, or an active mechanisms (hydroactive stomatal closure) induced as a metabolic response to an overall desiccation of leaves and roots: stomata close due to a reduced turgor of guard cells caused by a decline of the solute concentration. In the active process an important role is played by abscisic acid (ABA): ABA can reach the guard cells from the mesophyll, when it is dehydrated, or from the roots. The ABA synthesised in roots is transported to the shoot through the xylem; the ABA concentration in xylem sap is very low in normal condition, but under drought stress it increases dramatically. It is universally accepted that stomatal behaviour is strictly related to soil moisture, due to ABA produced in dehydrating roots; in fact, stomata closure is induced by soil drying, even if leaves water potential is maintained constant (Morgan 1990; Taylor 1991; Turner et al. 2001). Cytokinins can induce stomata opening directly or by reduction of the sensitivity of guard cells to ABA (Wilkinson and Davies 2002). ABA role under drought stress is also played in increasing the root-to-shoot ratio, inducing the plant to produce more and deeper roots and inhibiting the leaf area development.

As mentioned above, photosynthesis is impaired in drought-stressed plants by reduced leaf area and stomata closure; however direct negative effects of dehydration on photosynthesis pigments (Anjum et al. 2003) and enzymes have also been detected. The decrease in cell turgor potential can induce denaturation of enzymes or hinder their activities (Hoekstra et al. 2001). Under drought stress, Rubisco small subunits transcript abundance is also decreased (Vu et al. 1999); moreover, in these conditions Rubisco acts as oxygenase more than carboxylase, further reducing the efficiency of the process. Activities of phosphoenolpyruvate carboxylase, NADP-malic enzyme, fructose-1,6-biphosphatase, and pyruvate orthophosphate dikinase also decrease when leaves water potential lowers and ATP synthesis is reduced under drought stress (Farooq et al. 2009).

Reduced carbon dioxide influx can also cause an increased electron flux that results in reactive oxygen species (ROS) formation, which accumulate causing damages to cell membranes and proteins (Fu and Huang 2001; Farooq et al. 2009). Plants can apply several defence strategies against ROS through enzymatic or non-enzymatic antioxidant systems; among them can be listed beta-carotenes, alpha-tocopherols, ascorbate, reduced glutathione, and enzymes such as superoxide dismutase, peroxidise, ascorbate peroxidise, catalase, polyphenol oxidase and glutathione reductase (Farooq et al. 2009).

Plant cells must maintain a water potential lower than soil matric potential to be able to absorb water. At the permanent wilting point the soil water potential is lower than plant osmotic potential: plants can not maintain the turgor pressure even if transpiration is completely stopped. When soil dries, cells can increase solutes concentrations to decrease their water potential without losing turgor pressure or cell volume; this process is called osmotic adjustment.

Inorganic ions (potassium in particular) can be accumulated in vacuole, while in the cytoplasm the concentration of compatible solutes is increased. These are organic compounds that not interfere with cellular metabolism, such as proline, glycinebetaine, citrulline, gamma-aminobutyric acid, trehalose, dehydrins; these molecules can act simply as osmotic regulators or also by stabilizing cellular structures and molecules. Osmotic adjustment has been well studied in leaves, but it also happens in roots; the maintenance of turgor pressure in root meristems allows to maintain root growth (Farooq et al. 2009).

Water flow through the xylem is guaranteed by the negative hydrostatic pressure generated by the leaves and by the cohesive forces exerted by water into the xylematic vessels. During drying the plant resistance to water flow increases: roots shrinking due to desiccation causes damages to root hairs; root growth slows down and the root surface becomes covered of suberin, that makes it impermeable to water; moreover, cavitation, the breakage of water columns under tension within the xylem, can happen. As cavitation can more easily happen in large vessels, under drought stress plants produce vessels with smaller internal diameter, because in narrow columns the cohesive forces of water are higher. Reduced water flow is the main cause of reduced nutrient uptake from soil, that is considered the main cause of reduced growth and yield of plants due to water deficiency.

Morphological changes can be induced by drought to escape from sun and heat, such as leaves movements (paraheliotropic leaves, leaves wilting or rolling), pubescence (hairs that reflect radiation) or reflexive wax depositions.

## 4.1. Role of Rhizospheric Microorganisms in Drought Stress Alleviation

The main microbial populations known as active in alleviating plant drought stress are arbuscular mycorrhizal (AM) fungi and plant growth promoting rhizobacteria (PGPR)

### *4.1.1. Arbuscular Mycorrhizal Fungi (AMF)*

AMF form endosymbiotic associations with the roots of nearly all the terrestrial plants and their role in improving plant tolerance to drought stress has been observed long since. The mechanisms by which AMF exert their positive action are indirect or direct.

The main indirect mechanism is due to the improvement of the soil structure. Soil porosity is fundamental in determining the water content and the rate of water movement in soil, from which the root water uptake depends. The amount of moisture retained by a soil is called the field capacity; field capacity is high in clay soils or in soils reach in organic matter, while it is low in sandy soils, whose particle diameters are the biggest. Plants are less exposed to risk of drought stress when they grow in soils with high field capacity. However, the actual water availability to plant is also dependent from the soil water potential, that is determined by the surface tension and by the adhesive forces between water and soil particle surfaces, and by the soil hydraulic conductivity, that is a measure of the ease with which the water moves through the soil. In fact, when plant roots absorb water, the soil is depleted where it is in contact with the root surface; in the area of depletion a negative pressure gradient forms, that draws water from farer zones. Water moves in soil through bulk flow, that is possible if water channels are formed by the interconnected filled pores. Hydraulic conductivity falls when pores are filled with air and water channels are interrupted; that's why a soil with an unfavourable structure can contain a significant amount of water, that notwithstanding is not available to plant. AMF can affect soil water retention properties by forming and stabilizing soil aggregates, in particular in the size range of 20-200 μm diameter, through the entrapment of the soil particles by the hyphae. Moreover, extra-radical mycelial growth significantly improves the deposition of organic carbon in the rhizospheric soil, that, in addition, increases the growth and activities of other non-symbiotic organisms. The amounts of AMF carbon deposition, turnover rates and persistence as organic matter in soil are little known. Hyphal length density associated to roots is usually reported in an average value of 40 $m \cdot g^{-1}$(Hallet et al. 2009; Miller and Jastrow 2002). It was evaluated that the AMF spores, that can be considered as a persistent organic carbon form in soil, are present at 28 $spores \cdot g^{-1}$ soil, that correspond to 919 $kg \cdot ha^{-1}$ (Sieverding et al. 1989). Great attention was received by glomalin, a glycoprotein produced by AMF that is considered as an important source of soil organic matter and that contribute to the improvement of soil structure (Smith et al. 2010).

An increased plant water uptake due to a direct water transfer through AMF hyphae has been proposed, but the idea remains controversial. AMF grow in the roots and form extra-radical mycelium that can extend and explore a great volume of soil, also far beyond the roots. AMF hyphae can penetrate soil pores that are unaccessible to root hairs due to very small diameters: in fact, the medium diameter of a hypha is 2-5 μm, while smallest root hairs have a diameter of about 10 μm. Then, AM plants are able to uptake a greater volume of water from soil than non-AM plants. Several authors consider the contribution to water uptake of hyphae negligible for the plant (Fitter 1985; George et al. 1992; Koide 1993). However, a positive effect of AMF on plant growth and water uptake was demonstrated when water was delivered in a hyphal compartment of soil, where the penetration of roots was not allowed (Ruiz-Lozano and Azcon 1995).

AMF symbiosis also affects behaviour of plants stressed by drought acting on several physiological responses.

The hydraulic root conductivity seems to be affected by AM symbiosis. Hydraulic root conductivity is influenced by aquaporins, membrane water channels that allow the symplastic movement of water in roots. Even if drought stress seems not to induce a higher expression of plasma membrane aquaporin genes (PIP) in AM plants, the level of PIP2 protein was decreased in non-AM plants under water deficiency, while it remained constant in AM plants (Aroca et al. 2007).

AM and non-AM plants often show different transpiration rates and stomatal conductances, but in several studies no differences were found. In several studies it was reported that AM plants maintain higher gas exchange rates than non-AM plants when water availability is depleted in soil, even when nutrient availability was comparable. In AM plants stomatal closure appears at a leaf water potential lower than in non-AM plants, indicating different critical points or thresholds of stomatal behaviour during water depletion (Augé 2001). Higher stomatal conductance and transpiration rates were also observed in AM plants with shoot and leaves water content higher than non-AM plants (Ruiz-Lozano 2003). Higher leaf water content was also found associated to a lower ABA concentration in xylem sap, indicating a lower stress status of roots in drying soil (Duan et al. 1996). A delay in leaf water potential decline during drought stress and a more rapid return to control levels after relief of drought due to AM symbiosis was also observed (Augé 2001).

Fungal influence on stomatal opening is closely linked to growth promotion, under well-watered or water-depleted conditions (Augé 2001). In fact, stomatal opening favours higher photosynthetic rates, and increased transpiration favours higher nutrient uptake from soil, in particular phosphorous uptake (see above); moreover, a higher chlorophyll content was observed in AM plants, that remained higher after drought (Augé et al. 1987; Davies et al. 1993; Gemma et al. 1997). AM fungi can also affect the host balances of ABA and ABA/cytokinin (Augé 2001).

The effect of AM symbiosis on leaf osmotic adjustment was studied, with contrasting results. Leaf osmotic potential may differ in AM and non-AM plants during drought or under optimal water availability. In general, AM plants show higher osmotic potential of leaves than non-AM plants, indicating a higher tolerance to drought stress. An increase in proline content of AM plants under drought (Ruiz-Lozano et al. 1995; Azcon et al. 1996; Goioechea et al. 1998) and differences in free amino acids and sugars pools in roots (Augé et al. 1992) were reported.

Several enzymatic activities were found higher in AM plants during drought. Nitrate reductase, the first enzyme of the nitrogen assimilation pathway that is strongly reduced by several stress agents, is higher in AM plants and is also produced by AM fungi; this was proposed as a possible cause of preserved biomass production of AM plants under drought stress (Ruiz-Lozano 2003). Superoxide dismutases, antioxidant enzymes active against the reduced oxygen species (ROS), and glutathione reductase, a component of the ascorbate-glutathione cycle that regenerates oxidized glutathione into its reduced forms, are higher in AM plants (Ruiz-Lozano 2003).

#### *4.1.2. Plant Growth Promoting Rhizobacteria (PGPR)*

Several studies have been carried out to evaluate the role of PGPR in affecting the plant response to drought stress (Saleem et al. 2007; Kang et al. 2010; Yang et al. 2010).

Ethylene production by plants is one of the main consequences of exposure to several stress factors, including drought stress; in fact, ethylene regulates plant omeostasis under stress conditions reducing shoot and root growth. 1-aminocyclopropane-1-carboxylate (ACC) deaminase-producing PGPR promote the plant growth by reducing the ethylene concentration inside the plant (see above).

Several PGPR, such as *Paenibacillus polymyxa* (Timmusk et al. 1999), *Achromobacter piechaudii* (Mayak et al. 2004), and mixtures of PGPR-N fixing rhizobia (Figueiredo et al. 2008) or PGPR-AM fungi (Kohler et al. 2008), have been recognised as capable of alleviating the consequences of drought stress in plants. These studies evidenced in general an improved tolerance of plants to drought, with higher growth and yields, and synergistic effects from the use of mixtures were found. PGPR can also indirectly improve the plant response to drought stress by promoting root development, increased root surface area or increased number of root tips, due to the production of phytormones such as IAA (Yang et al. 2010). A study on *Azospirillum* effect on wheat growth enhancement has been recently carried out (Arzanesh et al. 2011). *Azospirillum* is able to survive and to promote plant growth and yield under drought, even if reduced soil moisture has a negative effect on soil microorganism and on the association efficiency between microorganisms and plants. In fact, soil drying affects bacterial physiological functions, mobility, chemotaxis and adsorption to soil particles; moreover, drought has an adverse effect on root exudates, that are necessary for the plant-microbes interaction. *Azospirillum*-wheat association induced a decrease of leaf water potential and an increase of leaf water content, that was attributed to the production of plant hormones by bacteria, such as ABA and IAA, that enhanced root growth and formation of lateral roots and, consequently, nutrients and water uptake (Arzanesh et al. 2011). Similar results were obtained by Casanovas et al. (2003) on *Zea mays* L. with *Azospirillum brasilense*. Sandhya et al. (2010) found that inoculation of maize seeds with

*Pseudomonas* spp. under drought improved plant biomass, relative water content, leaf water potential, roots adhering soil/root tissue ratio, soil aggregates stability and mean weight diameter and decreased leaf water loss. Moreover, plants showed a higher content of proline, sugars and free amino acids, but drought caused a lower content of proteins and starch. Inoculation with PGPR lowered the activity of antioxydant enzymes, such as ascorbate peroxidase, catalase and glutathione peroxidase, indicating that inoculated plants were less stressed.

#### *4.1.3. Maize Tolerance to Drought Stress and Inoculation of Specifically Selected Rhizospheric Microorganisms: A Case Study*

The utilization of microbial strains isolated from arid soils and selected for their tolerance to osmotic stress to increase the tolerance of plants toward reduced water availability is a strategy that was applied in a study with maize (Bardi et al. 2010).

A consortium of drought-tolerant rhizospheric microorganisms was produced as a mix of selected AMF and PGPR (Consortium 1). The microorganisms were selected from a rhizospheric soil sampled in Louga, a sahelian region of Senegal. AMF were selected and propagated according to Luster and Finlay (2006) with *Sorghum bicolor* as host plant. Two *Bacillus* spp. and two *Pseudomonas* spp. strains were selected under osmotic stress pressure on selective media. This consortium was compared to a commercial consortium (Consortium 2), non specific for drought tolerance (Micosat F, CCS Aosta s.r.l., Italy) composed of a mix of AMF (*Glomus caledonium* GM24, *Glomus intraradices* GG31, *Glomus coronatum* GU53, in form of spores, hyphae and root fragments) and PGPR (*Pseudomonas fluorescens* PA28, *Pseudomonas borealis* PA29, *Bacillus subtilis* BA41). The microbial inocula were used to inoculate maize plants at sowing near the seed (1 g · plant$^{-1}$). The trials were conducted in mesocosms in a glasshouse, to simulate a field assay, but under controlled condition that allowed to control the amount of water delivered to the plants, in order to induce drought stress; the control was formed by not inoculated plants. Each mesocosm was constituted of a plastic bag containing 1 m$^3$ of soil, and six maize plants were grown in it. The mesocosms were formed of two separated compartments, similarly to the experimental system proposed by Ruiz-Lozano and Azcon (1995): an upper compartment where roots could grow, and a lower compartment separated by a tissue allowing AM hyphae penetration but excluding roots. Plants were watered, from sowing until bloom, with an adequate water volume to restore field capacity, according to the watering planning reported in Table 1.

**Table 1. Water regimes of mesocosms planned to test the effect of rhisospheric selected microorganisms on maize response to drought stress**

| Water regimes | Upper compartment | Watering intervals | Lower compartment | Watering intervals |
|---|---|---|---|---|
| 1 | X | 1 week | X | 1 week |
| 2 | X | 1 week | - | |
| 3 | X | 2 weeks | - | |
| 4 | X | 2 weeks | X | 2 weeks (alternated to watering of upper compartment) |
| 5 | X | 2 weeks | X | 2 weeks (contemporary to watering of upper compartment) |

Measure of gas and water exchange parameters (stomatal conductance, transpiration rate, $CO_2$ uptake, internal $CO_2$, water use efficiency, leaf water content), and of chlorophyll index were carried out on plants under water deficiency (14 days after irrigation) and in the recovery phase (1 day after irrigation). Biomass yield (ears and shoots dry weight) was also determined at the end of the growth period.

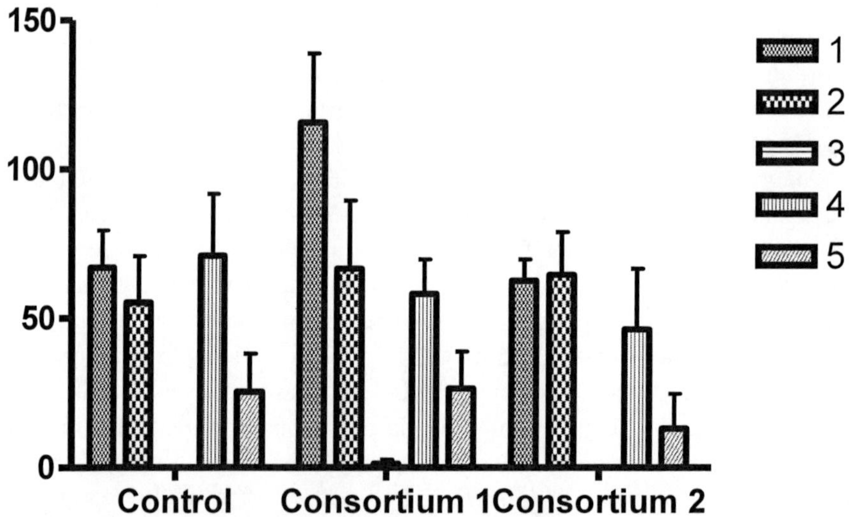


Figure 2. Ears biomass (dry weight) obtained at different water regimes (see Table 1). *Consortium 1*: Pool of AMF isolated from arid soil and PGPR selected for tolerance to osmotic stress. *Consortium 2*: Commercial non-specific consortium of rhizospheric microorganisms.

Plant yield and physiology were affected by the microbial consortia used for inoculation. Plants inoculated with the microbial consortium selected from the arid environment produced the highest biomass yield (shoots and ears, Figures 2 and 3), in particular when well-watered (treatment 1), and allowed to reach the production of ears with treatment 3, where the drought stress was the highest and where the plants not inoculated and also the plants inoculated with the commercial consortium did not form any ear. The shoot biomass was highest, when applying the consortium with selected drought-tolerant microorganisms, with the water regimes 4 and 5. These results suggest that this consortium improved the biomass yield under well-watered and drought-stressed conditions as well. The results of the water regimes 2 and 4 (Figure 3) confirm that water uptake from the compartment lacking root was improved by the consortium of microorganisms selected from the arid soil, suggesting that AM hyphae present in that compartment were able to improve water uptake of plants.

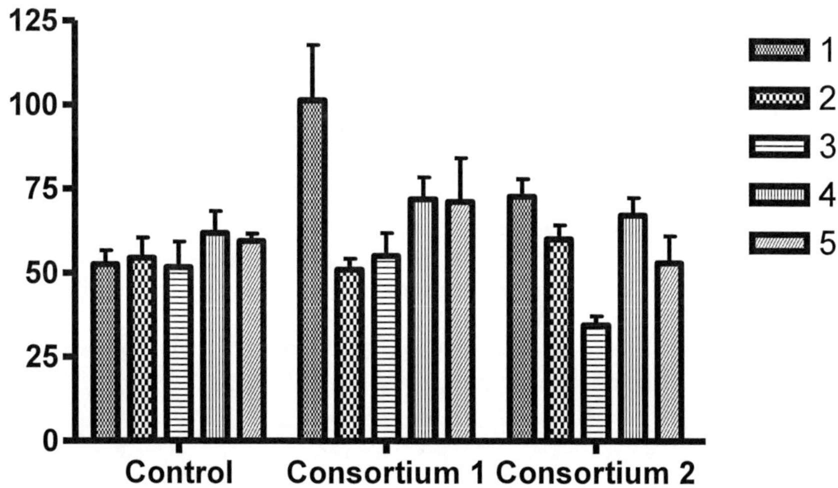


Figure 3. Shoot biomass (dry weight) obtained at different water regimes (see Table 1). *Consortium 1*: Pool of AMF isolated from arid soil and PGPR selected for tolerance to osmotic stress. *Consortium 2*: Commercial non-specific consortium of rhizospheric microorganisms.

The gas exchange parameters (stomatal conductance, transpiration rate, $CO_2$ uptake, internal $CO_2$) were improved by the specifically selected consortium both under water deficiency and during recovery phase (data not shown). The stomatal conductance ($g_s$), for example (Figure 4), pointed to a better hydration state of plants inoculated with both consortia, but with a better condition induced by the consortium formed by drought-tolerant microbial strains. Under water deficiency, the positive effect of this consortium was most evident in water regime 5, that correspond to the most prolonged time without watering (2 weeks, in both upper and lower compartment).

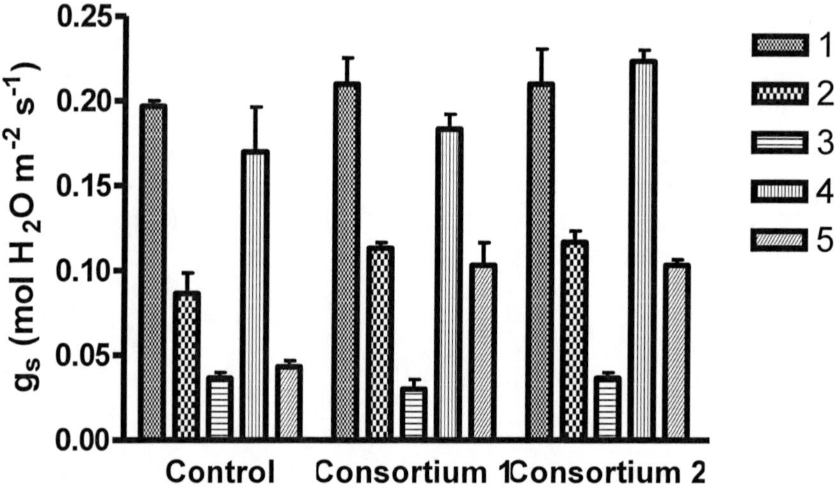


Figure 4. Stomatal conductance at different water regimes (see Table 1). *Consortium 1*: Pool of AMF isolated from arid soil and PGPR selected for tolerance to osmotic stress. *Consortium 2*: Commercial non-specific consortium of rhizospheric microorganisms.

However, no differences were found for chlorophyll index and water use efficiency between the different treatments (data not shown).

## 5. CONCLUSION

Since plants that are resistant to one stress are often more resistant to other stresses (Bowler and Fluhr 2000), the use of mycorrhizal fungi and PGPR to improve oxidative stress tolerance could be a promising strategy to protect plants from different abiotic stresses.

Application of PGPR containing ACC deaminase in agriculture might also increase the tolerance of plants to abiotic stresses and promoting more sustainable methods of crop production (Sturz et al. 2000). However, one of the major drawbacks in the large scale employment of plant growth-promoting bacteria is that these organisms may not always survive harsh environmental conditions. Nevertheless, several strains of symbiotic and non-symbiotic bacteria as well as mycorrhiza inocula are already marketed worldwide for this purposes (Burd et al. 2000; Cocking 2003; Jeffries et al. 2003) and the application of biofertilizers (containing PGPR and/or mycorrhizal fungi) is gaining momentum as a practice allowing the relief of nutrient stresses.

Nevertheless, there are a number of issues that have to be optimized when dealing with efficient inoculation techniques. The formulation of inocula that have a reliable and consistent effect under field conditions is still a bottleneck for their wider and reliable use. In this respect, the choice of the technology for inocula production and of the carrier for the formulation are key to their successful application. Any manipulation of rhizosphere biology to improve nutrient and/or water uptake shall be considered in the framework of the farming system in which microbial-based technologies are applied. Minimum tillage or precision agriculture are practices that are becoming increasingly utilized for annual crops, for which the application of inocula can be most easily implemented. A holistic approach should be taken into consideration also to benefit from other technologies (for example fertigation/irrigation systems, seed encapsulation, etc.). Indeed, the best results are expected to arise from synergistic interactions among many agronomical practices working together.

Identification of traits linked with the ability of selected microorganisms to establish in the rhizosphere and compete or cooperate with autochthonous microorganisms will be key for the development of technologies delivering beneficial microorganisms to plant crops. A possible solution to this problem may lie in the use of endophytic plant growth-promoting bacteria (Sturz and Nowak 2000) or in the selection of species-specific strains (Malusà and Sas Paszt 2009). Transformation of strains with genes involved in Pi solubilisation can also be a strategy to produced improved biofertilizers. Some have been cloned and their expression in not P-solubilising strains allowed the solubilisation of insoluble P minerals (Goldstein and Liu 1987). In this respect, the role of research is essential to link rhizosphere biology and crop productivity. Molecular methodologies have dominated the study of rhizosphere organisms so far. Nucleic acids can be extracted from soil microorganisms, and analysed to identify previously uncultured organisms or evaluate ecological diversity of communities (Marschner et al. 2001). Other traditional techniques, such as BIOLOG, or novel culturing methods (Janssen et al. 2002) are providing new insights of the microbiological diversity in the rhizosphere. Fluorescence *in situ* hybridisation (FISH) allows showing the positioning of bacteria on roots (Amann et al. 1995). Even though the technique presents some drawbacks, the information obtained could be further increased combining FISH with laser confocal microscopy. Another interesting tool helping understanding the interaction between roots and rhizosphere microorganisms can be that of reporter genes expressing fluorescing proteins in response to metabolic compounds related to rhizosphere activity (Jaeger et al. 1999). Finally, the development of image technologies is also opening new avenues for studying rhizosphere structure and processes. Cryo-analytical scanning electron microscopy (Ryan et al. 2003), computed tomography for noninvasive imaging of root–water–organism interactions (Johnson et al. 2004) and minirhizotrons are among the most advanced methods in this respect. However, to obtain the highest efficacy from the application of microbial technologies, it will be necessary to better understand the behaviour of rhizosphere microorganisms through time and in relation to different agronomical practices. Indeed, there is a limited knowledge about the length and velocity of a given rhizospheric process in relation to different agronomical practices.

In conclusion, the results achieved till now in utilizing microbial strains selected for a specific function or a specific plant/microorganism interaction are encouraging to apply these strategies. The improvement of plant response to drought or nutritional stresses when specifically selected microbial strains are used is a successful example. Nevertheless, more knowledge about the processes and the regulation of the plant-rhizosphere microorganisms interaction is needed to standardize a practice which is supporting the sustainability of agricultural productions.

## AKNOWLEDGMENTS

Some of the results shown in this work are part of projects financed from Piedmont Region, Directorate Agriculture, Sector Agricultural Development (Italy). The support to E.M. from the EU Regional Development Fund through the Polish Innovation Economy Operational Program, contract N. UDA-POIG.01.03.01-10-109/08-00 is also acknowledged.

Reviewed by Anna Benedetti, CRA-RPS Agricultural Research Council, Research Centre for the Soil-Plant System, Via della Navicella, 2/4, 00184 Rome (Italy), e-mail anna.benedetti@entecra.it

## References

[1] Abbate, P.E., Dardanellib,J.L., Cantareroc, M.G., Maturanoc, M., Melchiorid, R.J.M., Sueroa, E.E. (2004) Climatic and water availability effects on water use efficiency in wheat. *Crop Sci.*, 44, 474-483.

[2] Abd-Alla, M. H. (1994). Use of organic phosphorus by *Rhizobium leguminosarum* bv. viciae phosphatases. *Biol Fertil. Soils*, 8, 216–218.

[3] Adesemoye, A.O., Torbert, H. A., Kloepper, J. W. (2009). Plant Growth-Promoting Rhizobacteria Allow Reduced Application Rates of Chemical Fertilizers. *Microb Ecol,* 58, 921–929.

[4] Adesemoye, A.O., Kloepper, J.W. (2009). Plant-microbes interactions in enhanced fertilizer use efficiency. *Appl Microbiol Biotechnol*, 85, 1–12.

[5] Adesemoye, A.O., Torbert, H.A, Kloepper, J.W. (2008). Enhanced plant nutrient use efficiency with PGPR and AMF in an integrated nutrient management system. *Can J Microbiol*, 54, 876–886.

[6] Alguacil, M.M., Hernandez, J.A., Caravaca, F., Portillo, B., Roldan, A. (2003). Antioxidant enzyme activities in shoots from three mycorrhizal shrub species afforested in a degraded semi-arid soil. *Physiol Plant*, 118, 562–70.

[7] Alikhani, H.A., Saleh-Rastin, N., Antoun, H. (2006). Phosphate solubilization activity of rhizobia native to Iranian soils. *Plant Soil,* 287, 35–41.

[8] Altomare, C., Norvell, W. A., Bjorkman, T. and Harman, G. E. (1999). Solubilization of Phosphates and Micronutrients by the Plant-Growth-Promoting and Biocontrol Fungus *Trichoderma harzianum* Rifai 1295-22. *App Environ Microbiol*, 65, 2926–2933.

[9] Amann, R.I., Ludwig,W., Schleifer, K.H. (1995). Phylogenetic identification and *in situ* detection of individual microbial cells without cultivation. *Microbiological Reviews* 59, 143–169.

[10] Amora-Lazcano, E., Vazquez, M.M. and Azcon, R. (1998). Response of nitrogen-transforming microorganisms to arbuscular mycorrhizal fungi. *Biol Fertil Soils,* 27, 65– 70.

[11] Andrade, G., Mihara, K.L., Linderman, R.G., and Bethlenfalvay, G.J. (1997). Bacteria from rhizosphere and hyphosphere soils of different arbuscular-mycorrhizal fungi. *Plant Soil,* 192, 71–79.

[12] Anjum, F., Yaseen, M., Rasul, E., Wahid, A., Anjum, S. (2003) Water stress in barley (*Hordeum vulgare* L.). I. Effect on chemical composition and chlorophyll content, Pakistan *J. Agr. Sci.*, 40, 45-49.

[13] Antoun, H, A., Beauchamp, C. J., Goussard, N., Chabot, R. and Lalande, R. (1998). Potential of *Rhizobium* and *Bradyrhizobium* species as plant growth promoting rhizobacteria on nonlegumes: effect on radishes (*Raphanus sativus* L.). *Plant Soil,* 204, 57–67.

[14] Apel, K., Hirt, H. (2004). Reactive oxygen species: metabolism, oxidative stress, and signal transduction. *Annu Rev Plant Biol*, 55, 373–399.

[15] Aristizabal, C., Rivera, E.L., Janos, D.P. (2004). Arbuscular mycorrhizal fungi colonize decomposing leaves of *Myrica parvifolia, M. pubescens* and *Paepalanthus* sp.. *Mycorrhiza*, 14, 221–228.

[16] Aroca R, Porcel R, Ruiz-Lozano M (2007) How does arbuscular mycorrhizal symbiosis regulate root hydraulic properties and plasma membrane aquaporins in *Phaseolus vulgaris* under drought, cold and salinity stresses.? *New Phytol* 173, 808–816.

[17] Artursson, V., Finlay, R.D., and Jansson, J.K. (2005). Combined bromodeoxyuridine immunocapture and terminal restriction fragment length polymorphism analysis highlights differences in the active soil bacterial metagenome due to *Glomus mosseae* inoculation or plant species. *Environ Microbiol,* 7, 1952–1966.

[18] Arzanesh, M. H., • Alikhani, H. A., • Khavazi, K., • Rahimian, H. A.,• Miransari, M. (2011) Wheat (*Triticum aestivum* L.) growth enhancement by *Azospirillum* sp. under drought stress. *World J Microbiol Biotechnol*, 27, 197–205.

[19] Aseri, G.K., Jain, N., Panwar, J., Rao, A.V., Meghwal, P.R. (2008). Biofertilizers improve plant growth, fruit yield, nutrition, metabolism and rhizosphere enzyme activities of pomegranate (*Punica granatum* L.) in Indian Thar Desert. *Sci Hortic*, 117, 130–135.

[20] Atul-Nayyar, A., Hamel, C., Hanson, K., Germida, J. (2009). The arbuscular mycorrhizal symbiosis links N mineralization to plant demand. *Mycorrhiza*, 19, 239–246

[21] Augé, R.M. (2001). Water relations, drought and vesicular-arbuscular mycorrhizal Symbiosis. *Mycorrhiza*, 11, 3–42.

[22] Augé, R.M., Foster, J.G., Loescher, W.H., Stodola, A.W. (1992) Symplastic sugar and free amino acid molality of Rosa roots with regard to mycorrhizal colonization and drought. *Symbiosis* ,12, 1-17.

[23] Augé, R.M., Schekel, K.A., Wample, R. L. (1987) Leaf water and carbohydrate status of VA mycorrhizal rose exposed to drought stress. *Plant Soil,* 99,291-302.

[24] Azcón, R., Gomez, M., Tobar, R. (1992). Effects of nitrogen source on growth, nutrition, photosynthetic rate and nitrogen metabolism of

mycorrhizal and phosphorous-fertilized plants of *Lactuca sativa* L. *New Phytol*, 121, 227–234.

[25] Azcon, R., Gomez, M., Tobar, R.M. (1996) Physiological and nutritional responses by *Lactuca sativa* to nitrogen sources and mycorrhizal fungi under drought. *Biol Fertil Soils*, 22,156–161

[26] Azcón, R., Rodríguez, R., Amora-Lazcano, E., Ambrosano, E. (2008). Uptake and metabolism of nitrate in mycorrhizal plants as aVected by water availability and N concentration in soil. *Eur J Soil Sci*, 59, 131–138.

[27] Azcón, R., Ruiz-Lozano, J.M., Rodrıguez, R. (2001). Differential contribution of arbuscular mycorrhizal fungi to plant nitrate uptake (15N) under increasing N supply to the soil. *Can J Bot*, 79, 1175–1180.

[28] Azcón-Aguilar, C., Rodriguez-Navarro, D., Barea, J. (1981). Effects of ethrel on the formation and responses to VA mycorrhiza in *Medicago* and *Triticum*. *Plant Soil,* 60, 461–468.

[29] Baldani, V.L.D., Baldani, J.I., Dobereiner, J. (2000). Inoculation of rice plants with the endophytic diazotrophs *Herbaspirillums seropidicae*. *Biol Fertil Soils*, 30, 485–491.

[30] Bardi, L., Malusà, E., Zoppellari, F., Rosso, F., Sala, G.,Chitarra, W., Lovisolo, C. (2010) Effetto di inoculi microbicirizosferici selezionati sulla tolleranza del mais allo stress idrico: il progetto Idrostress. *Quaderni della Regione Piemonte – Agricoltura,* 73 (Suppl.: Annali del Settore Fitosanitario Regionale), 37-41.

[31] Barea, J.M., Azcon, R., and Azcon-Aguilar, C. (2002). Mycorrhizosphere interactions to improve plant fitness and soil quality. *Antonie Van Leeuwenhoek,* 81, 343–351.

[32] Barea, J.M., Navarro, E., Montoya, E. (1976). Production of plant growth regulators by rhizosphere phosphate solubilizing bacteria. *J Appl Bacteriol,* 40, 129–134.

[33] Barea, J.M., Pozo, M.J., Azcón, R., Azcón-Aguilar, C. (2005). Microbial co-operation in the rhizosphere. *J Exp Bot*, 56, 1761–1778.

[34] Barea, J.M., Toro, M., Orozco, M.O., Campos, E., Azcón, R. (2002). The application of isotopic (32P and 15N) dilution techniques to evaluate the interactive effect of phosphatesolubilizing rhizobacteria, mycorrhizal fungi and *Rhizobium* to improve the agronomic efficiency of rock phosphate for legume crops. *Nutr Cycl Agroecosys*, 63, 35–42.

[35] Bauer, W. D., Mathesius, U. (2004). Plant responses to bacterial quorum sensing signals. *Curr Opinion Plant Biol*, 7, 429–433.

[36] Benedito, V.A., Li, H., Dai, X., Wandrey, M., He, J., Kaundal, R., Torres-Jerez, I., Gomez, S.K., Harrison, M.J., Tang, Y., Zhao, P.X. (2010).

Genomic inventory and transcriptional analysis of *Medicago truncatula* transporters. *Plant Physiol*, 152, 1716–1730.

[37] Bianciotto, V., Lumini, E., Lanfranco, L., Minerdi, D., Bonfante, P. and Perotto, S. (2000). Detection and identification of bacterial endosymbionts in arbuscular mycorrhizal fungi belonging to the family *Gigasporaceae*. *App Environ Microbiol*, 66, 4503–4509.

[38] Bieleski, R.L. (1973). Phosphate pools, phosphate transport and phosphate availability. *Annu Rev Plant Physiol*, 24, 225–252.

[39] Biró, B., Köves-Péchy, K., Vörös, I., Takács, T., Eggenberger, P., and Strasser, R.J. (2000). Interrelations between *Azospirillum* and *Rhizobium* nitrogen fixers and arbuscular mycorrhizal fungi in the rhizosphere of alfalfa in sterile AMFfree or normal conditions. *Appl Soil Ecol*, 15, 159–168.

[40] Blaha, D., Prigent-Combaret, C., Mirza, M.S., Moenne-Loccoz, Y. (2006) Phylogeny of the 1-aminocyclopropane-1-carboxylic acid deaminase-encoding gene acdS in phytobeneficial and pathogenic *Proteobacteria* and relation with strain biogeography. *FEMS Microbiol Ecol*, 56, 455–470

[41] Boddey, R.M.,Urquiaga, S., Reis, V., Döbereiner, J. (1991). Biological nitrogen fixation associated with sugar cane. *Plant Soil*, 137, 111–117

[42] Bonfante, P. and Requena, N. (2011). Dating in the dark: how roots respond to fungal signals to establish arbuscular mycorrhizal symbiosis. *Curr. Opinion Plant Biol*, 14, 451-457

[43] Bonkowski, M. (2004). Protozoa and plant growth: the microbial loop in soil revisited *New Phytol,* 162, 617–631

[44] Bonkowski, M., Cheng, W., Griffith, B.S., Alphei, J., Scheu, S. (2000). Microbial-faunal interactions in the rhizosphere and effects on plant growth *Eur. J. Soil Biol*, 36, 135−147

[45] Bowler, C, Fluhr, R. (2000). The role of calcium and activated oxygen as signals for controlling cross-tolerance *Trends Plant Sci*, 5, 241–246.

[46] Bray EA (1997) Plant responses to water deficit. *Trends Plant Sci*, 2, 48–54.

[47] Bressano, M., Curetti, M., Giachero, L., Vargas Gil, S., Cabello, M., March, G., Ducasse, D.A., Luna, C.M. (2010). Mycorrhizal fungi symbiosis as a strategy against oxidative stress in soybean plants. *J Plant Physiol*, 167, 1622–1626.

[48] Bucher, M. (2006). Functional biology of plant phosphate uptake at root and mycorrhiza interfaces. *New Phytol*, 173, 11–26.

[49] Burd, G., Dixon, D.G., Glick, B.R. (2000). Plant growth promoting bacteria that decrease heavy metal toxicity in plants. *Can J Microbiol*, 46, 237–245.

[50] Burleigh, S.H., Cavagnaro, T.R., Jakobsen, I. (2002). Functional diversity of arbuscular mycorrhizas extends to the expression of plant genes involved in P nutrition. *J Exp Bot*, 53, 1–9.

[51] Cappellazzo , G., Lanfranco, L., Fitz, M., Wipf, D. and Bonfante, P. (2008). Characterization of an amino acid permease from the endomycorrhizal fungus *Glomus mosseae* . *Plant Physiol*, 147, 429 – 437.

[52] Cardoso, I.M., Boddington, C.L., Janssen, B.H., Oenema, O., Kuyper, T.W. (2006). Differential access to phosphorus pools of an oxisol by mycorrhizal and nonmycorrhizal maize. *Comm Soil Sci Plant Anal*, 37, 1537–1551.

[53] Caris , C., Hördt, W., Hawkins, H.J., Römheld, V., George, E. (1998). Studies of iron transport by arbuscular mycorrhizal hyphae from soil to peanut and sorghum plants *Mycorrhiza*, 8, 35–39.

[54] Cavagnaro, T.R., Smith, F.A., Smith, S.E., Jakobsen, I. (2005). Functional diversity in arbuscular mycorrhizas: exploitation of soil patches with different phosphate enrichment differs among fungal species. *Plant Cell Environ*, 28, 642–650.

[55] Cesco, S., Neumann, G., Tomasi, N., Pinton, R., Weisskopf, L. (2010). Release of plant-borne flavonoids into the rhizosphere and their role in plant nutrition. *Plant Soil*, 329, 1–25.

[56] Chabot, R., Antoun, H., Kloepper, J.W. and Beauchamp, C.J. (1996). Root colonization of maize and lettuce by bioluminescent *Rhizobium leguminosarum* biovar phaseoli. *Appl Environ Microbiol*, 62, 2767–2772.

[57] Cheng, Z., Park, E., Glick, B.R. (2007). 1-Aminocyclopropane-1-carboxylate deaminase from *Pseudomonas putida* UW4 facilitates the growth of canola in the presence of salt. *Can J Microbiol*, 53, 912–918.

[58] Christensen, H., and Jakobsen, I. (1993). Reduction of bacterial growth by a vesicular-arbuscular mycorrhizal fungus in the rhizosphere of cucumber (Cucumis sativus L.). *Biol Fert Soils*, 15, 253–258.

[59] Clark, R.B., Zeto, S.K. (1996a). Iron acquisition by mycorrhizal maize grown on alkaline soil. *J Plant Nutr*, 19, 247–264.

[60] Clark, R.B., Zeto, S.K. (1996b). Mineral acquisition by mycorrhizal maize grown on acid and alkaline soil. *Soil Biol Biochem*, 28, 1495–1503.

[61] Clark, R.B., Zeto, S.K. (2000). Mineral acquisition by arbuscular mycorrhizal plants. *J Plant Nutr*, 23, 867–902.

[62] Cocking. E.C. (2003). Endophytic colonization of plant roots by nitrogen-fixing bacteria. *Plant Soil*, 252, 169–175.

[63] Davies, F.T., Potter, J.R., Linderman, R.G. (1993) Drought resistance of mycorrhizal pepper plants independent of leaf P-concentration – response in gas exchange and water relations. *Physiol Plant,* 87, 45-53

[64] DeLucia, E.H., Heckathorn, S.A. (1989) The effect of soil drought on water-use efficiency in a contrasting Great Basin desert and Sierran montane species, *Plant Cell Environ.*, 12, 935-940.

[65] Dey, R., Pal, K.K., Bhatt, D.M., Chauhan, S.M. (2004). Growth promotion and yield enhancement of peanut (*Arachis hypogaea* L.) by application of plant growth-promoting rhizobacteria. *Microbiol Res*, 159, 371–94.

[66] Dixon, R.A. and Paiva, N.L. (1995). Stress-induced phenylpropanoid metabolism. *Plant Cell,* 7, 1085-1097.

[67] Dobbelaere, S., Vanderleyden, J., Okon, Y. (2003). Plant growth-promoting effects of diazotrophs in the rhizosphere. *Critical Reviews in Plant Sciences*, 22, 107–149.

[68] Drew, E.A., Murray, R.S., Smith, S.E. and Jakobsen, I. (2003). Beyond the rhizosphere: growth and function of arbuscular mycorrhizal external hyphae in sands of varying pore sizes. *Plant Soil,* 251, 105–114.

[69] Duan, X., Neuman, D.S., Reiber, J.M., Green, C.D., Saxton, A.M., Augé, R.M. (1996). Mycorrhizal influence on hydraulic and hormonal factors implicated in the control of stomatal conductance during drought. *Journal of Experimental Botany,* 47, 1541-1550.

[70] Duff, S.M.G., Sarath, G., and Plaxton, W.C. (1994). The role of acid phosphatases in plant phosphorus metabolism. *Physiol Plant,* 90, 791-800.

[71] Dugassa, G., von Alten, H., Schonbeck, F. (1996). Effects of arbuscular mycorrhiza (AM) on health of *Linum usitatissimum* L. infected by fungal pathogens. *Plant Soil,* 185, 173–182.

[72] Duijff, B.J., De Kogel, W.J., Bakker, P.A.H.M., Schippers, B. (1994). Influence of pseudobactin 358 on the iron nutrition of barley. *Soil Biol Biochem*, 26, 1681–1994.

[73] Egamberdiyevaa, D., Höflich, G. (2004). Effect of plant growth-promoting bacteria on growth and nutrient uptake of cotton and pea in a semi-arid region of Uzbekistan *J Arid Environ,* 56, 293–301.

[74] Ezawa, T., and Yoshida, T. (1994). Characterization of phosphatase in marigold roots infected with vesicular-arbuscular mycorrhizal fungi. *Soil Sci. Plant Nutri,* 40, 255-264.

[75] Ezawa, T., Hayatsu, M., and Saito M. (2005). A new hypothesis on the strategy for acquisition of phosphorus in arbuscular mycorrhiza: up-regulation of secreted acid phosphatase gene in the host plant. *Mol Plant Microbe Interact,* 18, 1046–1053.

[76] Faraldo-Gomez, J.D., Sansom, M.S. (2003). Acquisition of siderophores in Gram-negative bacteria. *Nat Rev Mol Cell Biol,* 4, 105–116.

[77] Farooq, M., Wahid, A., Kobayashi, N., Fujita, D., and Basra, S.M.A. (2009) Plant drought stress: Effects, Mechanisms and Management. In: *E. Lichtfouse* et al. (eds.), Sustainable Agriculture, SpringerScience+Business Media B.V. – EDP Sience, pp 153-188.

[78] Figueiredo, V.B. et al. (2008) Alleviation of drought stress in the common bean (*Phaseolus vulgaris* L.) by co-inoculation with *Paenibacillus polymyxa* and *Rhizobium tropici*. *Appl. Soil Ecol,* 40, 182-188.

[79] Fitter, A.H. (1985) Functioning of vesicular-arbuscular mycorrhizas under field conditions. *New Phytol*, 99, 257–265

[80] Franche, C., Lindström, K., Elmerich, C. (2009). Nitrogen-fixing bacteria associated with leguminous and non-leguminous plants. *Plant Soil*, 321, 35-59

[81] Fu, J., Huang, B. (2001) Involvement of antioxidants and lipid peroxidation in the adaptation of two cool-season grasses to localized drought stress. *Environ. Exp. Bot.,* 45, 105-114.

[82] Gamalero, E., Martinotti, M.G., Trotta, A., Lemanceau, P., and Berta, G. (2004). Morphogenetic modifications induced by *Pseudomonas fluorescens* A6RI and *Glomus mosseae* BEG12 in the root system of tomato differ according to plant growth conditions. *New Phytol*, 155, 293–300.

[83] Gange, A. (2000). Arbuscular mycorrhizal fungi, collembola and plant growth. *Trends Ecol Evol*, 15, 369–372.

[84] Geil, R., Guinel, F. (2002). Effects of elevated substrate-ethylene on colonization of leek (*Allium porrum*) by the arbuscular mycorrhizal fungus *Glomus aggregatum*. *Can J Bot*, 80, 114–119.

[85] George, E., Hauser, K.U., Vetterlein, D., Gorgus, E., Marschner, H. (1992) Water and nutrient translocation by hyphae of *Glomus mosseae*. *Can J Bot,* 70, 2130–2137.

[86] George, E., Häussler, K-U., Vetterlein, D., Gorgus, E., Marschner, H. (1992). Water and nutrient translocation by hyphae of *Glomus mosseae*. *Can J Bot*, 70, 2130–2137.

[87] Giri, B., Giang, P.H., Kumari, R., Prasad, R., Varma, A. (2005). Microbial diversity in soils. In F. Buscot, A. Varma (Eds.), *Microorganisms in soils: Roles in genesis and functions.* (pp 195–212). Heidelberg, D: Springer-Verlag.

[88] Glassop, D., Smith, S.E., Smith, F.W. (2005). Cereal phosphate transporters associated with the mycorrhizal pathway of phosphate uptake into roots. *Planta*, 222, 688–698.

[89] Glick, B.R. (2005). Modulation of plant ethylene levels by the bacterial enzyme ACC deaminase. *FEMS Microbiol Lett*, 251, 1–7.

[90] Glick, B.R., Cheng, Z., Czarny, J., Duan, J. (2007). Promotion of plant growth by ACC deaminase-producing soil bacteria. *Eur J Plant Pathol*, 119, 329–339.

[91] Glick, B.R., Penrose, D.M., Li, J. (1998). A model for the lowering of plant ethylene concentrations by plant growth-promoting bacteria. *J Theor Biol*, 190, 63–68.

[92] Goicoechea, N., Szalai, G., Antolin, M.C., Sanchez-Diaz, M., Paldi, E. (1998) Influence of arbuscular mycorrhizae and *Rhizobium* on free polyamines and proline levels in water-stressed alfalfa. *J Plant Physiol*, 153, 706–711.

[93] Goldstein, A.H., Liu, S.T. (1987). Molecular cloning and regulation of a mineral phosphate solubilizing gene from *Erwinia herbicola*. *Biotechnology*, 5, 72–74.

[94] Govindarajulu, M., Pfeffer, P.E., Hairu, J., Abubaker, J., Douds, D.D., Allen, J.W., Bucking, H., Lammers, P.J., Shachar-Hill, Y. (2005). Nitrogen transfer in the arbuscular mycorrhizal symbiosis. *Nature*, 435, 819–823.

[95] Gryndler, M. (2000). Interactions of arbuscular mycorrhizal fungi with other soil organisms. In Y., Kapulnik, and D.D.L. Douds (Eds.*), Arbuscular mycorrhizas: physiology and function* (pp. 239–262). Dordrecht, the Netherlands: Kluwer Academic Publishers.

[96] Guerinot, M.L. (1994). Microbial iron transport. *Annu Rev Microbiol*, 48, 743–772.

[97] Guether, M., Neuhäuser, B., Balestrini, R., Dynowski, M , Ludewig, U. and Bonfante, P. (2009). A mycorrhizal-specific ammonium transporter from *Lotus japonicus* acquires nitrogen released by arbuscular mycorrhizal fungi. *Plant Physiol*, 150,73 – 83.

[98] Guo, Y., George, E., Marschner, H. (1996). Contribution of an arbuscular mycorrhizal fungus to the uptake of cadmium and nickel in bean and maize plants. *Plant Soil*, 184, 195–205.

[99] Gyaneshwar, P., Naresh Kumar, G., Parekh, L.J., Poole, P.S. (2002). Role of soil microorganisms in improving P nutrition of plants. *Plant Soil,* 245, 83–93.

[100] Hallet, P.D., Feeney, D.S., Bengough, A.G. et al (2009) Disentangling the impact of AM fungi versus roots on soil structure and water transport. *Plant Soil*, 314,183–196.

[101] Hamel, C. (2004). Impact of arbuscular mycorrhizal fungi on N and P cycling in the root zone. *Can J Soil Sci,* 84, 383–395.

[102] Hammond,, J.P., Bennett, M.J., Bowen, H.C., Broadley, M.R., Eastwood, D.C., May, S.T., Rahn, C., Swarup, R., Woolaway, K.E., and White, P.I.

(2003). Changes in gene expression in Arabidopsis shoots during phosphate starvation and the potential for developing smart plants. *Plant Physiol,* 132, 578–96.

[103] Hartmann, A., Rothballer, M., Schmid, M. (2008). Lorenz Hiltner, a pioneer in rhizosphere microbial ecology and soil bacteriology research. *Plant Soil* 312:7–14

[104] Hartmann, A., Schmid, M., van Tuinen, D., Berg, G. (2009). Plant-driven selection of microbes. *Plant Soil,* 321, 235–257.

[105] Hause, B., Mrosk, C., Isayenkov, S., Strack D. (2007). Jasmonates in arbuscular mycorrhizal interactions. *Phytochemistry*, 68, 101–110.

[106] Hawkins, H.J., Johansen, A., George, E. (2000). Uptake and transport of organic and inorganic nitrogen by arbuscular mycorrhizal fungi. *Plant Soil,* 226, 275–285.

[107] Hayat, R., Ali, S., Amara, U., Khalid, R., Ahmed, I. (2010). Soil beneficial bacteria and their role in plant growth promotion: a review. *Ann Microbiol,* 60, 579–598.

[108] Hodge, A. (2003). Plant nitrogen capture from organic matter as affected by spatial dispersion, interspecific competition and mycorrhizal colonization. *New Phytol*, 157, 303–314.

[109] Hodge, A. (2004). The plastic plant: root responses to heterogeneous supplies of nutrients. *New Phytol*, 162, 9–24.

[110] Hodge, A., Campbell, C.D., Fitter, A.H. (2001). An arbuscular mycorrhizal fungus accelerates decomposition and acquires nitrogen directly from organic material. *Nature*, 413, 297–299.

[111] Hodge, A., Fitter, A.H. (2010). Substantial nitrogen acquisition by arbuscular mycorrhizal fungi from organic material has implications for N cycling. *Proc Natl Acad Sci USA*, 107, 13754–13759.

[112] Hoekstra, F.A., Golovina, E.A., Buitink, J. (2001) Mechanisms of plant dessication tolerance, *Trends Plant Sci.*, 6, 431-438.

[113] Honma, M., Shimomura, T. (1978). Metabolism of 1-amino-cyclopropane-1-carboxylic acid. *Agric Biol Chem*, 42, 1825–1831.

[114] Howell, C.R. (2003). Mechanisms employed by *Trichoderma* species in the biological control of plant diseases: the history and evolution of current concepts *Plant Disease*, 87, 4-10.

[115] Hurek, T., Reinhold-Hurek, B., Van Montagu, M., Kellenberger, E. (1994). Root colonization and systemic spreading of *Azoarcus* sp. strain BH72 in grasses. *J Bacteriol,* 176, 1913–1923.

[116] Illmer, P., Schinner, F. (1995). Solubilization of inorganic calcium phosphates-solubilization mechanisms. *Soil Biol Biochem*, 27, 257–263.

[117] Ishii, T., Shrestha, Y., Matsumoto, I., Kadoya, K., (1996). Effect of ethylene on the growth of vesicular-arbuscular mycorrhizal fungi and on the mycorrhizal formation of trifoliate orange roots. *J Japan Soc Hort Sci*, 65, 525–529.

[118] Jaeger, C., Lindow, S., Miller, S., Clark, E., Firestone, M. (1999). Mapping of sugar and amino acid availability in soil around roots with bacterial sensors of sucrose and tryptophan. *Applied and Environmental Microbiology* 65, 2685–2690.

[119] Janssen, P.H., Yates, P.S., Grinton, B.E., Taylor, P.M., Sait, M. (2002). Improved culturability of soil bacteria and isolation in pure culture of novel members of the divisions *Acidobacteria*, *Actinobacteria*, *Proteobacteria* and *Verrucomicrobia. Applied and Environmental Microbiology* 68, 2391–2396.

[120] Jayasinghearachchi, H.S., Seneviratne, G. (2006). Fungal solubilization of rock phosphate is enhanced by forming fungal–rhizobial biofilms. *Soil Biol Biochem*, 38, 405–408.

[121] Jeffries, P., Gianinazzi, S., Perotto, S., Turnau, K., Barea, J.M. (2003). The contribution of arbuscular mycorrhizal fungi in sustainable maintenance of plant health and soil fertility. *Biol. Fertil. Soils*, 37, 1-16.

[122] Jetiyanon, K., Kloepper, J.W. (2002). Mixtures of plant growth promoting rhizobacteria for induction of systemic resistance against multiple plant diseases. *Biol Control*, 24, 285–291.

[123] Johansen, A., Jakobsen, I., Jensen, E.S. (1993). Hyphal transport by a vesicular- arbuscular mycorrhizal fungus of N applied to the soil as ammonium or nitrate. *Biol Fertil Soils*, 16, 66–70.

[124] Johansson, J.F., Paul, L.R., Finlay, R.D. (2004). Microbial interactions in the mycorrhizosphere and their significance for sustainable agriculture. *FEMS Microb Ecol*, 48, 1-13.

[125] Johnson, S.N., Read, D.B., Gregory, P.J. (2004). Tracking larval insect movement within soil using high resolution X-ray microtomography. *Ecological Entomology* 29, 117–122.

[126] Joner, E.J., van Aarle, I.M., Vosatka, M. (2000). Phosphatase activity of extra-radical arbuscular mycorrhizal hyphae. *Plant Soil,* 226, 199–210.

[127] Jones, D.L., Dennis, P.G., Owen, A.G., van Hees, P.A.W. (2003). Organic acid behavior in soils—misconceptions and knowledge gaps. *Plant Soil,* 248, 31–41.

[128] Jones, D.L., Owen, A.G., Farrar, J.F. (2002). Simple method to enable the high resolution determination of total free amino acids in soil solutions and soil extracts. *Soil Biol Biochem*, 34, 1893–1902.

[129] Jungk, A. (2001). Root hairs and the acquisition of plant nutrients from soil. *J Plant Nutr Soil Sci*, 164, 121–129.

[130] Juszczuk, I.M., Malusà, E., and Rychter, A.M. (2001). Oxidative stress during phosphate deficiency in roots of bean plants (*Phaseolus vulgaris* L.). *J Plant Physiol*, 158, 1299-1305.

[131] Kang, B. G., Kim, W. T., Yun, H. S., Chang, S. C. (2010) Use of plant growth-promoting rhizobacteria to control stress response of plant root. *Plant Biotechnol.* Rep, 4, 179-183.

[132] Karandashov, V., and Bucher, M. (2005). Symbiotic phosphate transport in arbuscular mycorrhizas. *Trends Plant Sci*, 10, 22–29.

[133] Kennedy, A.C. (1998) The rhizosphere and spermosphere. In: D.M. Sylvia, J.J. Fuhrmann, P.G. Hartel and D.A. Zuberer (eds), *Principles and applications of soil microbiology*, Prentice-Hall International (UK) Limited, London, 389-407.

[134] Kim, K.Y., Jordan, D., and McDonald, G.A. (1998). Effect of phosphate-solubilizing bacteria and vesicular-arbuscular mycorrhizae on tomato growth and soil microbial activity. *Biol Fert Soils*, 26, 79–87.

[135] Kloepper, J.W. (1996). Host specificity in microbe-microbe interactions. *Bioscience*, 46, 406–409.

[136] Kobae, Y., Tamura, Y., Takai, S., Banba, M., and Hata, S. (2010). Localized expression of arbuscular mycorrhiza-inducible ammonium transporters in soybean. *Plant Cell Physiol*, 51, 1411–1415.

[137] Kohler, J. et al. (2008) Plant-growth-promoting rhizobacteria and arbuscular mycorrhizal fungi modify alleviation biochemical mechanisms in water-stressed plants. *Funct. Plant Biol.*, 35, 141–151

[138] Koide, R. (1993) Physiology of the mycorrhizal plant. *Adv Plant Pathol*, 9, 33–54

[139] Kuzyakov, Y. (2002). Review: Factors affecting rhizosphere priming effects. *J Plant Nutr Soil Sci*, 165, 382–396.

[140] Lambert, D.H., Baker, D.E., Cole, H.Jr (1979). The role of mycorrhizae in the interactions of phosphorus with zinc, copper, and other elements. *Soil Sci Soc Am J*, 43, 976–980.

[141] Lazaridou M., Kirilov, A., Noitsakis, B., Todorov, N., Katerov, I. (2003) the effect of water deficit on yield and water use efficiency of lucerne. Optimal forage systems for animal production and the environment. In: *Proceedings of the 12th Symposium of the European Grassland Federation*, Pleven, Bulgaria, 26-28 May 2003.

[142] Leigh , J. , Hodge, A., and Fitter, A.H. (2009). Arbuscular mycorrhizal fungi can transfer substantial amounts of nitrogen to their host plant from organic material . *New Phytol*, 181, 199–207.

[143] Lemanceau, P., Bauer, P., Kraemer, S., Briat, J-F. (2009). Iron dynamics in the rhizosphere as a case study for analyzing interactions between soils, plants and microbes *Plant Soil*, 321, 513–535.

[144] Li, S.M., Li, L., Zhang, F.S. (2004). Enhancing phosphorus and nitrogen uptake of faba bean by inoculating arbuscular mycorrhizal fungus and *Rhizobium* leguminosarum. *J China Agri Uni,* 9, 11–15.

[145] Lisette, J., Xavier, C., Germida, J.J. (2003). Selective interactions between arbuscular mycorrhizal fungi and *Rhizobium leguminosarum* bv. *viceae* enhance pea yield and nutrition. *Biol Fertil Soils*, 37, 261–267.

[146] Liu, A., Plenchette, C., Hamel, C. (2007). Soil nutrient and water providers: how arbuscular mycorrhizal mycelia support plant performance in a resource-limited world. In: C. Hamel, C. Plenchette (Eds.), *Mycorrhizae in crop production* (pp 38–66). Binghamton, GB: Haworth.

[147] Liu, H., Trieu, A. T., Blaylock, L. A., and Harrison, M. J. (1998). Cloning and characterization of two phosphate transporters from *Medicago truncatula* roots: Regulation in response to phosphate and to colonization by arbuscular mycorrhizal (AM) fungi. *Mol Plant-Microbe Interact*, 11, 14-22.

[148] Lopez-Bucio, J., Cruz-Ramirez, A., Herrera-Estrella, L. (2003). The role of nutrient availability in regulating root architecture. *Curr Opin Plant Biol*, 6, 280–87.

[149] Lopez-Bucio, J., Hernandez-Abreu, E., Sanchez-Calderon, L., Nieto-Jacobo, M.F., Simpson, J., Herrera-Estrella, L. (2002). Phosphate availability alters architecture and causes changes in hormone sensitivity in the *Arabidopsis* root system. *Plant Physiol*, 129, 244–56.

[150] Lucy, M., Reed, E., Glick, B.R. (2004). *Application of free living plant growth-promoting rhizobacteria*. Antonie van Leeuwenhoek, 86, 1–25.

[151] Ludwig-Muller, J. (2007) Indole-3-butyric acid synthesis in ecotypes andmutants of *Arabidopsis thaliana* under different growth conditions. *J. Plant Physiol.*, 164, 47-59.

[152] Luster, J., Finlay, R. (2006). *Handbook of Methods Used in Rhizosphere Research*. Birmensdorf, Swiss Federal Research Institute WSL. p 536.

[153] Ma, W., Charles, T. C., and Glick, B. R. (2004). Expression of an exogenous 1-aminocyclopropane-1-carboxylate deaminase gene in *Sinorhizobium meliloti* increases its ability to nodulate alfalfa. *Applied and Environmental Microbiology*, 70, 5891–5897.

[154] Ma, W., Guinel, F.C., Glick, B.R. (2003). *Rhizobium leguminosarum* biovar *viciae* 1-aminocyclopropane-1-carboxylate deaminase promotes nodulation of pea plants. *Appl Environ Microbiol*, 69, 4396–4402.

[155] Malik, K.A., Mirza, M.S., Hassan, U., Mehnaz, S., Rasul, G., Haurat, J., Bauy, R., Normanel, P. (2002). The role of plant associated beneficial bacteria in rice-wheat Cropping System. In: I.R. Kennedy, A.T.M.A. Chaudhry (Eds.) Biofertilisers in action. *Rural industries research and development* (pp 73–83). Canberra, AU: Corporation.

[156] Malusà, E., Laurenti, E., Juszczuk, I., Ferrari, R.P., Rychter, A.M. (2002). Free radical production in roots of *Phaseolus vulgaris* subjected to phosphate deficiency stress. *Plant Physiol Biochem*, 40, 963-967.

[157] Malusà, E., Russo, M.A., Mozzetti, C. and Belligno, A. (2006). Modification of secondary metabolism and flavonoid biosynthesis under phosphate deficiency in bean (*Phaseolus vulgaris* l.) roots. *J Plant Nutrition,* 29, 1-14.

[158] Malusà, E. and Sas Paszt, L. (2009). The development of innovative technologies and products for organic fruit production. An integrated project. *The Proceedings of the International Plant Nutrition Colloquium XVI.* Sacramento (USA). Paper 1359. http://repositories.cdlib.org/ipnc/xvi/1359.

[159] Malusà, E., Sas Paszt, L., Trzcinski, P., Górska, A. (2011). Influences of different organic fertilizers and amendments on nematode trophic groups and soil microbial communities during strawberry growth. *Acta Horticulturae*, in print

[160] Malusà, E., Sas-Paszt, L., and Zurawicz, E. (2007). The effect of a mycorrhiza-bacteria substrate and foliar fertilization on growth response and rhizosphere pH of three strawberry cultivars. *Intern J Fruit Science*, 6, 25-41.

[161] Marschner, H., Dell, B. (1994). Nutrient uptake in mycorrhizal symbiosis. *Plant Soil*, 59, 89–102.

[162] Marschner, H., Römheld, V. (1994). Strategies of plants for acquisition of iron. *Plant Soil,* 165, 261–274.

[163] Marschner, P., Crowley, D., Lieberei, R. (2001). Arbuscular mycorrhizal infection changes the bacterial 16 S rDNA community composition in the rhizosphere of maize. *Mycorrhiza*, 11, 297–302.

[164] Marschner, P., Yang, C.H., Lieberei, R., Crowley, D.E. (2001). Soil and plant specific effects on bacterial community composition in the rhizosphere. *Soil Biology and Biochemistry* 33, 1437–1445.

[165] Masalha, J., Kosegarten, H., Elmaci, Ö., Mengel, K. (2000). The central role of microbial activity for iron acquisition in maize and sunflower. *Biol Fert Soils*, 30, 433-439.

[166] Mayak, S., Tirosh, T., Glick, B.R. (2004). Plant growth-promoting bacteria that confer resistance to water stress in tomatoes and peppers. *Plant Sci,* 166, 525–530.

[167] Meyer, J.M., Abdallah, M.A. (1978). The fluorescent pigment of Pseudomonas fluorescens: biosynthesis, purification and physico-chemical properties. *J Gen Microbiol,* 107, 319–328.

[168] Miller, R.M., Jastrow, J.D. (2002) Mycorrhizal influence on soil structure. In: Kapulnik Y, Douds DD (eds) *Arbuscular mycorrhizae: molecular biology and physiology*. Kluwer Academic Publishers, Dordrecht, The Netherlands, pp 3–18

[169] Minerdi, D., Fani, R., Gallo, R., Boarino, A., and Bonfante, P. (2001). Nitrogen fixation genes in an endosymbiotic *Burkholderia* strain. *Applied and Environmental Microbiology*, 67, 725–732.

[170] Mittler, R. (2002). Oxidative stress, antioxidants and stress tolerance. *Trends Plant Science*, 7, 405-410.

[171] Morgan, P.W. (1990) Effects of abiotic stresses on plant hormone systems. In: *Stress responses in plants: adaptation and acclimation mechanisms.* Wiley-Liss, New York, pp. 113-146.

[172] Morgan, P.W., Drew, C.D. (1997). Ethylene and plant responses to stress. *Physiol Plant*, 100, 620–630.

[173] Mortimer, P.E., Pérez-Fernández, M.A., Valentine, A.J. (2011). Arbuscular mycorrhiza maintains nodule function during external NH4+ supply in *Phaseolus vulgaris* (L.). *Mycorrhiza*, DOI 10.1007/s00572-011-0396-9.

[174] Nacry, P., Canivenc, G., Muller, B., Azmi, A., Van Onckelen, H., Rossignol, M., and Doumas, P. (2005). A role for auxin redistribution in the responses of the root system architecture to phosphate starvation in Arabidopsis. *Plant Physiol*, 138, 2061–74.

[175] Okazaki, S., Nukui, N., Sugawara, M., Minamisawa, K. (2004). Rhizobial strategies to enhance symbiotic interactions: rhizobitoxine and 1-aminocyclopropane-1-carboxylate deaminase. *Microbes Environ,* 19, 99–111.

[176] Owen, A.G., Jones, D.L. (2001). Competition for amino acids between wheat roots and rhizosphere microorganisms and the role of amino acids in plant N acquisition. *Soil Biol Biochem*, 33, 651–657.

[177] Pacovsky, R.S. (1986). Micronutrient uptake and distribution in mycorrhizal and phosphorus fertilized soybeans. *Plant Soil*, 95, 379–385.

[178] Pacovsky, R.S., Fuller, G. (1988). Mineral and lipid composition of *Glycine–Glomus–Bradyrhizobium* symbioses. *Physiol Plant*, 72, 733–746.

[179] Panwar, J.D.S., Singh, O. (2000). Response of *Azospirillum* and *Bacillus* on growth and yield of wheat under field conditions. *Indian J Plant Physiol*, 5, 108–110.

[180] Parmar, N., Dadarwal, K.R. (1999). Stimulation of nitrogen fixation and induction of flavonoid-like compounds by rhizobacteria. *J Appl Microbiol,* 86, 36–44.

[181] Paszkowski, U., Kroken, S., Roux, C., Briggs S.P.(2002). Rice phosphate transporters include an evolutionarily divergent gene specifically activated in arbuscular mycorrhizal symbiosis. *Proc Natl Acad Sci USA*, 99, 13324–13329.

[182] Patten, C.L., Glick, B.R. (1996). Bacterial biosynthesis of indole-3- acetic acid. *Can J Microbiol*, 42, 207–220.

[183] Plessner, O., Klapach, T., Guerinot, M.L. (1993). Siderophore utilization by *Bradyrhizobium japonicum*. *Appl Environ Microbiol*, 59, 1688– 1690.

[184] Poulsen, K.H., Nagy, R., Gao, L.L., Smith S.E., Bucher, M., Smith, F.A., and Jakobsen, I. (2005). Physiological and molecular evidence for Pi uptake via the symbiotic pathway in a reduced mycorrhizal colonization mutant in tomato associated with a compatible fungus. *New Phytol*, 168, 445–453.

[185] Pozo, M.J., Loon, L.C.V., Pieterse, C.M.J., (2004). Jasmonates – signals in plant–microbe interactions. *J. Plant Growth Regul*, 23, 211–222.

[186] Raju, P.S., Clark, R.B., Ellis, J.R., Maranville, J.W. (1990). Effects of species of VA-mycorrhizal fungi on growth and mineral uptake of sorghum at different temperatures. *Plant Soil*, 121, 165–170.

[187] Revillas, J.J., Rodelas, B., Pozo, C., Martinez-Toledo, M.V., Gonzalez, L.J. (2000). Production of B-group vitamins by two *Azotobacter* strains with phenolic compounds as sole carbon source under diazotrophic and adiazotrophic conditions. *J Appl Microbiol*, 89, 486–493.

[188] Richardson, A.E. (2001). Prospects for using soil microorganisms to improve the acquisition of phosphorus by plants. *Aust J Plant Physiol*, 28, 897–906.

[189] Richardson, A.E., Barea, J.M., McNeill, A.M., Prigent-Combaret, C. (2009). Acquisition of phosphorus and nitrogen in the rhizosphere and plant growth promotion by microorganisms. *Plant Soil*, 321, 305–339.

[190] Richardson, A.E., George, T.S., Hens, M., Simpson, R.J. (2005). Utilization of soil organic phosphorus by higher plants. In: B.L. Turner, E. Frossard, D.S. Baldwin (Eds.) *Organic phosphorus in the environment* (pp 165–184). Wallingford, UK: CABI.

[191] Rillig, M.C., Mummey, D.L., Ramsey, P.W., Klironomos, J.N., Gannon, J.E. (2006). Phylogeny of arbuscular mycorrhizal fungi predicts community composition of symbiosis associated bacteria. *FEMS Microbiol Ecol*, 57, 389–395.

[192] Rodríguez, H., Fraga, R. (1999). Phosphate solubilizing bacteria and their role in plant growth promotion. *Biotechnol Adv,* 17, 319–339.

[193] Ron-Vaz, M.D., Edwards, A.C., Shand, C.A., Cresser, M.S. (1993) Phosphorus fractions in soil solution: influence of soil acidity and fertiliser additions. *Plant Soil,* 148, 175–183.

[194] Ruiz-Lozano, J.M. (2003). Arbuscular mycorrhizal symbiosis and alleviation of osmotic stress. New perspectives for molecular studies. *Mycorrhiza*, 13, 309-317.

[195] Ruiz-Lozano, J.M., Azcon, R., Gomez, M. (1995) Effects of arbuscular mycorrhizal *Glomus* species on drought tolerance: physiological and nutritional plant responses. *Appl Environ Microbiol*, 61, 456–460

[196] Ryan, M.H., McCully, M.E., Huang, C.X. (2003). Location and quantification of phosphorus and other elements in fully-hydrated, soil-grown arbuscular mycorrhizas: a cryo-analytical scanning electron microscopy study. *New Phytologist* 160, 429–441.

[197] Ryan, P. R, Delhaize, E., and Jones, D. L. (2001). Function and mechanism of organic anion exudation from plant roots. *Annu Rev Plant Physiol Plant Mol Biol*, 52, 527-560.

[198] Saleem, M., Arshad, M., Hussain, S., Bhatti, A.S. (2007). Perspective of plant growth promoting rhizobacteria (PGPR) containing ACC deaminase in stress agriculture *J Ind Microbiol Biotechnol*, 34, 635–648.

[199] Sánchez, E., Soto, J.M., Garcia, P.C., López-Lefebre, L.R., Rivero, R.M., Ruiz, J. M., and Romero, L. (2000) Phenolic and oxidative metabolism as bioindicators of nitrogen deficiency in French bean plants (*Phaseolus vulgaris* L. cv Strike). *Plant Biol,* 2, 272-277.

[200] Sandhya, V.,• Ali, Sk. Z.,• Grover, M.,• Reddy, G.,• Venkateswarlu, B. (2010) Effect of plant growth promoting Pseudomonas spp. on compatible solutes, antioxidant status and plant growth of maize under drought stress. *Plant Growth Regul*., 62, 21–30.

[201] Saubidet, M.I., Fatta, N., Barneix, A.J. (2002). The effect of inoculation with *Azospirillum brasilense* on growth and nitrogen utilization by wheat plants. *Plant Soil,* 245, 215–222.

[202] Scandalios, J.G. (1993). Oxygen stress and superoxide dismutases. *Plant Physiol*, 101, 7-12.

[203] Schachtman, D.P., and Shin, R. (2007). Nutrient sensing and signaling: *NPKS Annu Rev Plant Biol*, 58, 47-69.

[204] Schimel, J.P., Bennett, J. (2004). Nitrogen mineralization: Challenges of a changing paradigm. *Ecology*, 85, 591–602.

[205] Schimel, J.P., Weintraub, M.N. (2003). The implications of exoenzyme activity on microbial carbon and nitrogen limitation in soil: a theoretical model. *Soil Biol Biochem*, 35, 549–563.

[206] Schnepf, A., Roose, T., Schweiger, P. (2008). Impact of growth and uptake patterns of arbuscular mycorrhizal fungi on plant phosphorus uptake—a modelling study. *Plant Soil*, 312, 85–99.

[207] Secilia, J., and Bagyaraj, D.J. (1987). Bacteria and actinomycetes associated with pot cultures of vesicular arbuscular mycorrhizas. *Can J Microbiol*, 33, 1069–1073.

[208] Seefeldt, L.C., Dance, I.G., Dean, D.R. (2004). Substrate interactions with nitrogenase: Fe versus Mo. *Biochemistry*, 43, 1401-1409.

[209] Seeling, B., Zasoski, R.J. (1993). Microbial effects in maintaining organic and inorganic solution phosphorus concentrations in a grassland topsoil. *Plant Soil*, 148, 277–284.

[210] Sessitsch, A., Howieson, J.G., Perret, X., Antoun, H., Martínez- Romero, E. (2002). Advances in Rhizobium research. *Crit Rev Plant Sci*, 21, 323–378.

[211] Shaharoona, B., Arshad, M., Zahir, Z.A. (2006b). Performance of *Pseudomonas* spp. containing ACC-deaminase for improving growth and yield of maize (*Zea mays* L.) in the presence of nitrogenous fertilizer. *Soil Biol Biochem,* 38, 2971–2975.

[212] Shaharoona, B., Naveed, M., Arshad, M., Zahir, Z.A. (2008). Fertilizer dependent efficiency of *Pseudomonads* for improving growth, yield, and nutrient use efficiency of wheat (*Triticum aestivum* L.). *Appl Microbiol Biotechnol,* 79, 147–155.

[213] Shin, R., Berg, R.H., Schachtman, D.P. (2005). Reactive oxygen species and root hairs in *Arabidopsis* root response to nitrogen, phosphorus and potassium deficiency. *Plant Cell Physiol*, 46, 1350–57.

[214] Shin, R., Schachtman, D.P. (2004). Hydrogen peroxide mediates plant root response to nutrient deprivation. *Proc. Natl. Acad. Sci. USA*, 101, 8827–8832.

[215] Sieverding, E., Toro, S., Mosquera, O. (1989) Biomass production and nutrient concentrations in spores of VA mycorrhizal fungi. *Soil Biol Biochem,* 21, 69–72.

[216] Sivaramaiah, N., Malik, D.K., Sindhu, S.S. (2007). Improvement in symbiotic efficiency of chickpea (*Cicer arietinum)* by coinoculation of

*Bacillus* strains with *Mesorhizobium* sp. *Cicer*. *Indian J Microbiol*, 47, 51–56.

[217] Smith, F. W. (2002a). The phosphate uptake mechanism. *Plant Soil*, 245, 105- 114.

[218] Smith, F.A., Jakobsen, I., Smith, S.E. (2000). Spatial differences in acquisition of soil phosphate between two arbuscular mycorrhizal fungi in symbiosis with *Medicago truncatula. New Phytol*, 147, 357-366.

[219] Smith, S.E. (2002b) Soil microbes and plants—raising interest, mutual gains. *New Phytol*, 156, 142–144.

[220] Smith, S.E., Facelli, F., Pope, S., Smith, F.A.(2010). Plant performance in stressful environments: interpreting new and established knowledge of the roles of arbuscular mycorrhizas. *Plant Soil*, 326, 3–20.

[221] Smith, S.E., Read. D.J. (2008). Mycorrhizal Symbiosis, 3rd edn. New York, London, Burlington, San Diego: Elsevier and Academic.

[222] Smith, S.E., Smith, F.A., Jakobsen, I. (2003). Mycorrhizal fungi can dominate phosphate supply to plants irrespective of growth responses. Plant Physiology, 133, 16–20.

[223] Smith, S.E., Smith, F.A., Jakobsen, I. (2004). Functional diversity in arbuscular mycorrhizal (AM) symbioses: the contribution of the mycorrhizal P uptake pathway is not correlated with mycorrhizal responses in growth or total P uptake. *New Phytol*, 162, 511–524.

[224] Stajner, D., Kevreaan, S., Gasaić, O., Mimica-Dudić, N., Zongli, H. (1997). Nitrogen and *Azotobacter chroococcum* enhance oxidative stress tolerance in sugar beet. *Biol Plant*, 39, 441–445.

[225] Sturz, A. V., Nowak, J. (2000). Endophytic communities of rhizobacteria and the strategies required to create yield enhancing associations with crops. *Applied Soil Ecology*, 15, 183–190.

[226] Sturz, A.V., Christie, B.R., Novak, J. (2000). Bacterial endophytes: potential role in developing sustainable system of crop production. *Crit Rev Plant Sci,* 19, 1–30.

[227] Subramanian, K., Charest, C. (1999). Acquisition of N by external hyphae of an arbuscular mycorrhizal fungus and its impact on physiological responses in maize under drought-stressed and well-watered conditions. *Mycorrhiza*, 9, 69–75.

[228] Tan, R.X., Zou, W.X. (2001). Endophytes: A rich source of functional metabolites. *Nat Prod Rep*, 18, 448–459.

[229] Tawaraya, K., Naito, M., Wagatsuma, T. (2006). Solubilization of insoluble inorganic phosphate by hyphal exudates of arbuscular mycorrhizal fungi. *J Plant Nutr*, 29, 657–665.

[230] Taylor, I.B. (1991) Genetics of ABA synthesis. In: Davies, W.J., H.G. Jones (Eds.),*Abscisic acid: Physiology and Biochemistry*, BIOS Scientific, London, pp. 23-38.

[231] Timmusk, S. and Wagner, G.H. (1999) The plant-growth-promoting rhizobacterium *Paenibacillus polymyxa* induces changes in *Arabidopsis thaliana* gene expression: a possible connection between biotic and abiotic stress responses. *Mol. Plant Microbe Interact.*, 12, 951–959

[232] Tobar, R.M., Azcon, R., Barea, J.M. (1994a). The improvement of plant N acquisition from an ammonium-treated, drought-stressed soil by the fungal symbiont in arbuscular mycorrhizae. *Mycorrhiza*, 4, 105–108.

[233] Toljander, J.F., Lindahl, B.D., Paul, L.R., Elfstrand, M., Finlay, R.D. (2007). Influence of arbuscular mycorrhizal mycelial exudates on soil bacterial growth and community structure. *FEMS Microbiol Ecol*, 61, 295–304.

[234] Toro, M., Azcón, R., and Barea, J.M. (1997). Improvement of arbuscular mycorrhiza development by inoculation of soil with phosphate-solubilizing rhizobacteria to improve rock phosphate bioavailability (32P) and nutrient cycling. *Appl Environ Microbiol,* 63, 4408–4412.

[235] Toro, M., Azcón, R., and Barea, J.M. (1998). The use of isotopic dilution techniques to evaluate the interactive effects of *Rhizobium* genotype, mycorrhizal fungi, phosphate-solubilizing rhizobacteria and rock phosphate on nitrogen and phosphorus acquisition by *Medicago sativa*. New Phytol, 138, 265–273.

[236] Tran Vân, V., Berge, O., Ke, S.N., Balandreau, J., Heulin, T. (2000). Repeated beneficial effects of rice inoculation with a strain of *Burkholderia vietnamiensis* on early and late yield components in low fertility sulphate acid soils of Vietnam. *Plant Soil,* 218, 273–284.

[237] Tu, C., Bookerw, F. L., Watson, D.M, Chen, X., Rufty, T.M., Shiz, W., and Hu, S. J. (2006) Mycorrhizal mediation of plant N acquisition and residue decomposition: Impact of mineral N inputs. *Global Change Biology*, 12, 793–803.

[238] Tuomela, M., Vikman, M., Hatakka, A., Itävaara, M. (2000). Biodegradation of lignin in a compost environment: a review. *Biores Technol*, 72, 169-183.

[239] Turner, N.C., Wright, G.C., Siddique, K.H.M. (2001) Adaptation of grain legumes (pulses) to water-limited environments.*Adv. Agron.*, 71, 123-231.

[240] Valentine, A.J., Osbourne, B.A., Mitchell, D.T. (2002). Form of inorganic nitrogen influences mycorrhizal colonization and photosynthesis of cucumber. *Sci Hortic*, 92, 229–239.

[241] van der Heijden, M.G.A., Wiemken, A., Sanders, I.R. (2003). Different arbuscular mycorrhizal fungi alter coexistence and resource distribution between co-occurring plant. *New Phytol*, 157, 569–578.

[242] Vansuyt, G., Robin, A., Briat, J.F., Curie, C., Lemanceau, P. (2007). Iron acquisition from Fe-pyoverdine by *Arabidopsis thaliana*. *Mol Plant Microbe Interact*, 4, 441–447.

[243] Vázquez, M.M., Barea, J.M., Azcón, R. (2001). Impact of soil nitrogen concentration on *Glomus* spp.-*Sinorhizobium* interactions as affecting growth, nitrate reductase activity and protein content of *Medicago sativa*. *Biol Fertil Soils*, 34, 57–63.

[244] Vessey, J. K. (2003). Plant growth promoting rhizobacteria as biofertilizers. *Plant Soil,* 255, 571–586.

[245] Vierheilig, H., Alt, M., Mohr, U., Boller, T., Wiemken, A., (1994). Ethylene biosynthesis and activities of chitinase and □-1,3-glucanase in the roots of host and non-host plants of vesicular-arbuscular mycorrhizal fungi after inoculation with *Glomus mosseae*. *J Plant Physiol*, 143, 337–343.

[246] Vierheilig, H., Piche´,Y., (2002). Signalling in arbuscular mycorrhiza: facts and hypothesis. In: B. Buslig, J. Manthey (Eds.), *Flavonoids in cell function* (pp. 23–39). New York, USA: Kluwer Academic Publishers/PlenumPress.

[247] Visca, P., Imperi, F., Lamont, I.L. (2007). Pyoverdine siderophores: from biogenesis to biosignificance. *Trends Microbiol*, 15, 22–30.

[248] Vu, J.C.V., Gesch, R.W., Allen, L.H., Boote, K.J., Bowes, G. (1999) CO2 enrichment delays a rapid, drought-induced decrease in Rubisco small subunit transcript abundance. *J. Plant Physiol.*, 155, 139-142.

[249] Wakelin, S.A., Gupta, V.V.S.R., Harvey, P.R., Ryder, M.H. (2007). The effect of *Penicillium* fungi on plant growth and phosphorus mobilization in neutral to alkaline soils from southern Australia. *Can J Microbiol*, 53, 106–115.

[250] Walter, A., Römheld, V., Marschner, H., Crowley, D.E. (1994). Iron nutrition of cucumber and maize: effect of *Pseudomonas putida* YC 3 and its siderophore. *Soil Biol Biochem*, 26, 1023–1031.

[251] Wang, X., Pan, Q., Chen, F., Yan, X., Liao, H. (2011). Effects of co-inoculation with arbuscular mycorrhizal fungi and rhizobia on soybean growth as related to root architecture and availability of N and P. *Mycorrhiza*, 21, 173–181.

[252] Wasternack, C., Hause, B., (2002). Jasmonates and octadecanoids – signals in plant stress response and development. In: K. Moldave (Ed.), *Progress in Nucleic Acid Research and Molecular Biology* (pp. 165–221). New York, USA: Academic Press.

[253] Werner, D.D. (2001). Organic signals between plants and microorganisms. In *The rhizosphere*, R. Pinton, Z. Varanini, P. Nannipieri (Eds.) (pp 197-222). New York, USA: Marcel Dekker Inc.

[254] Whitehead, N. A., Barnard, A. M., Slater, H., Simpson, N. J., and Salmond, G. P. (2001). Quorum-sensing in Gram-negative bacteria. *FEMS Microbiology Reviews*, 25, 365–404.

[255] Whitelaw, M.A. (2000). Growth promotion of plants inoculated with phosphate-solubilizing fungi. *Adv Agron*, 69, 99–151.

[256] Whitelaw, M.A., Harden, T.J., Helayer, K.R. (1999). Phosphate solubilization in solution culture by the soil fungus *Penicillium radicum*. *Soil Biol Biochem*, 31, 655–665.

[257] Wilkinson, S., Davies,W.J. (2002) ABA-based chemical signalling: the coordination of responses to stress in plants. *Plant Cell Environ.*, 25, 195-210.

[258] Wu, Q.S., Zou, Y.N., Xia, R.X. (2006). Effects of water stress and arbuscular mycorrhizal fungi on reactive oxygen metabolism and antioxidant production by citrus (*Citrus tangerine*) roots *Eur J Soil Biol*, 42, 166–72.

[259] Xie, Z., Staehelin, C., Vierheili, H., Wiemken, A., Jabbouri, S., Broughton, W.J., Vogeli-Lange, R., Boller, T., Xie, Z.P. (1995). Rhizobial nodulation factors stimulate mycorrhizal colonization of undulating and nonnodulating soybeans. *Plant Physiol*, 108, 1519–1525.

[260] Yang, J., Kloepper, J. W. and Ryu, C.-M. (2009) Rhizosphere bacteria help plants tolerate abiotic stress. *Trends in Plant Science*, 14(1), 1-4.

[261] Yang, J., Kloepper, J.W., and Ryu, C-M. (2009). Rhizosphere bacteria help plants tolerate abiotic stress. *Trends in Plant Science*, 14, 1-4.

[262] Yanni, Y.G., Rizk, R.Y., Abd El-Fattah, F.K., Squartini, A., Corich, V., Giacomini, A., de Bruijn, F., Rademaker, J., Maya-Flores, J., Ostrom, P., Vega-Hernandez, M., Hollingsworth, R.I., Martinez- Molina, E., Ninke, K., Philip-Hollingsworth, S., Mateos, P.F., Velasquez, E., Triplett, E., Umali-Garcia, M., Anarna, J.A., Rolfe, B.G., Ladha, J.K., Hill, J., Mujoo, R., Ng, P.K., Dazzo, F.B. (2001). The beneficial plant growth-promoting association of *Rhizobium leguminosarum* bv. *trifolii* with rice roots. *Aust J Plant Physiol*, 28, 845–870.

[263] Zak, D.R., Pregitzer. K.S., King, J.S. (2000). Elevated atmospheric CO2, fine roots and the response of soil microorganisms: a review and hypothesis. *New Phytol*, 147, 201–222.

[264] Zehnder, G. W., Murphy, J. F., Sikora, E. J., Kloepper, J. W. (2001). Application of rhizobacteria for induced resistance. *European Journal of Plant Pathology,* 107, 39-50.

In: Abiotic Stress: New Research
Editors: Nikhil Haryana and Shreya Punj
ISBN: 978-1- 61942-197-4


*Chapter2*

# NADP-DEPENDENT ENZYMES AND ABIOTIC STRESS

***Veronika Doubnerová* and Helena Ryšlavá***
Department of Biochemistry, Faculty of Natural Science,
Charles University in Prague, Hlavova 2030, Czech Republic

## ABSTRACT

Abiotic stress often negatively affects photosynthesis, which is the main source of NADPH. Reduced form of the coenzyme represents an indispensable compound for biosynthetic reactions, antioxidant systems, and for enzymes involved in regulation. As the ratio $NADP^+$/NADPH in plants influences redox reactions, the control of cell redox homeostasis is important for balancing metabolic processes and redox-dependent signalling.

NADPH is produced mainly by enzymes from oxidative pentose phosphate pathway (glucose-6-phosphate dehydrogenase, 6-phosphogluconate dehydrogenase) but also other enzymes (e.g. NADP-isocitrate dehydrogenase, NADP-malic enzyme, NADP-glutamate dehydrogenase, and non-phosphorylating NADP-glyceraldehyde-3-phosphate dehydrogenase) represent an alternative source of NADPH in stressed plants. NADP-glutamate dehydrogenase is involved in catabolism of proteins and amino acids and thus provides NADPH. However, the role of this enzyme could be different under stress condition (e.g. in the case of increased protein synthesis). Reaction catalyzed with non-phosphorylating NADP-glyceraldehyde-3-phosphate dehydrogenase drives NADPH/$NADP^+$ ratio to biosynthetic processes.

Enzymes of antioxidant system, glutathione reductase and monodehydroascorbate reductase, on the other hand, represent enzymes utilizing NADPH that are involved in stress. These enzymes participate in scavenging of reactive oxygen species that are produced in early stages of oxidative stress incidence by NADPH-oxidase. Similarly, NADPH is essential for biosynthetic processes, e.g. for biosynthesis of osmotically active compounds that reduce effect of cold, drought, and salt stress ($\Delta^1$-pyrroline-5-carboxylate synthetase and $\Delta^1$-pyrroline-5-carboxylate reductase synthesizing proline, mannose-6-phosphate reductase forming mannitol). Enzymes participating in biosynthesis of isoprenoids (e.g. 3-hydroxy-3-methyl-glutaryl-CoA reductase) could be useful for the production of defence compounds (terpens, sterols, and phytohormones). Shikimate dehydrogenase is the key enzyme in shikimate pathway, which contributes to the synthesis of aromatic amino acids (phenylalanine, tyrosine, and tryptophan) flavonoids, chalcons, lignols, lignans, and lignins; all these compounds seem to play an important role under stress conditions. Last but not least is NADPH-dependent thioredoxin reductase, regulatory enzyme that affects numerous cell processes and oxidative-stress response.

## Abbreviations

G6PDH, glucose-6-phosphate dehydrogenase; 6PGDH, 6-phosphogluconate dehydrogenase; GR, glutathione reductase; HMGR, 3-hydroxy-3-methyl-glutaryl-CoA-reductase; M6PR, mannose-6-phosphate reductase; MDHAR, monodehydroascorbate reductase; NADP-ICDH, NADP-isocitrate dehydrogenase; NADP-GDH; NADP-glutamate dehydrogenase; NADP-ME, NADP-malic enzyme; NP-GAPDH, non-phosphorylating NADP-glyceraldehyde-3-phosphate dehydrogenase; NTR, NADP-thioredoxin reductase; 5PCS, $\Delta^1$-pyrroline-5-carboxylate synthetase; 5PCR, $\Delta^1$-pyrroline-5-carboxylate reductase; ROS, reactive oxygen species; SDH, shikimate dehydrogenase.

## 1. NADPH

Continuous supply of reducing equivalents in the form of NADPH by photosynthesis is crucial for plant growth, development, and cell homeostasis maintenance. NADPH is not required only for biosynthetic reaction, such as biosynthesis of fatty acids, amino acids, isoprenoids, phytohormones, and defence compounds, but it is also an indispensable compound for antioxidant and regulation systems. It is needed for NADPH-oxidase generating ROS, which act against stress in early stage of defence response (e.g. ROS induce hypersensitive reaction and systemic acquired resistance) [1]. However ROS also negatively affect many enzymes and cell structures and therefore these compounds have to be scavenged by antioxidant system. NADPH is involved in ascorbate-glutathione cycle (Halliwell-Asada cycle), in which it is coenzyme of two key antioxidant enzymes. Above mentioned processes are carefully regulated by NADPH availability [2]. The term redox state is commonly used to describe the balance of $NADH/NAD^+$ and $NADPH/NADP^+$ in the cell. The $NADPH/NADP^+$ ratio in plants is usually kept high, in favour of the reduced form [2]. Strict control of cellular redox homeostasis is essential for balancing metabolic processes and redox-dependent signalling [3]. In addition to glutathione-ascorbate system, also thioredoxin systems act as redox regulators of the cell [4]. In most organisms, thioredoxin is reduced by enzyme reactions with participations of NADPH [5].

It is evident that production of NADPH is necessary for the maintenances of normal metabolism, but what is the impact of limited NADPH availability caused by abiotic stress? It is known, that photosynthesis, the main source of NADPH in plants, is negatively affected by abiotic stress (e.g. by drought [6, 7], by chilling stress [8], or by salt irrigation [9] etc. Alternative source of NADPH beside photosynthesis are NADP-producing enzymes, especially enzymes of oxidative pentose phosphate pathway, but also other NADP-enzymes. This chapter is focused on the relationship between abiotic stress and NADP-dependent enzymes producing NADPH and enzymes that utilize NADPH.

## 2. Enzymes Producing NADPH and Abiotic Stress

### 2.1. Glucose-6-Phosphate Dehydrogenase and 6-Phosphogluconate Dehydrogenase

Glucose-6-phosphate dehydrogenase (EC 1.1.1.49, G6PDH) and 6-phosphogluconate dehydrogenase (EC 1.1.1.44, 6PGDH) are the key enzymes of the oxidative pentose phosphate (OPP) cycle. G6PDH is the first and the rate limiting enzyme of this pathway, catalyzing oxidation of D-glucose-6-phosphate to 6-phospho-D-gluconolactone with the concomitant formation of NADPH from $NADP^+$. Also the next enzyme of OPP pathway 6PGDH catalyzes the reaction (oxidative decarboxylation of 6-phosphogluconate to ribulose-5-phosphate), in which NADPH is generated (Figure 1). G6PDH and 6PGDH are widely distributed enzymes; in plant tissues they are located in cytosol, plastids, and peroxisomes [10-13]. The cytosolic isoenzyme of G6PDH is inhibited by

NADPH; the ratio NADPH/$NADP^+$ is determinant for the enzyme activity. The regulation of the plastid isoenzyme is more complex. This isoenzyme is deactivated by light, by the photoreduced thioredoxin m [1]. Function of G6PDH in cytosol consists in the production of NADPH for anabolic biosynthetic and assimilation processes similarly as in plastids, where G6PDH provides intermediates of shikimate pathway and nucleic acid biosynthesis [14]. G6PDH has been for a long time considered as an important component in the biosynthetic pathway for the formation of the sugar moiety of nucleic acids [15]. It has been shown, that G6PDH is involved in nitrogen assimilation [16].

Findings that G6PDH activity is essential to protect cells against even mild oxidative stress were firstly reported in the case of mammals [17]. In plants G6PDH is important for the activation of resistance to environmental stresses [18]; this enzyme was found to be involved in salt stress [16,18-21] cold stress [22], hyperhydricity [23], and metal toxicity stress [24,25].

The G6PDH gene belongs to five early salt-responding genes of wheat that became activated after exposure to salt stress [20]. The 6PGDH gene was also up-regulated by salt stress in rice [12]. Under stress condition G6PDH and 6PGDH as key enzymes of pentose-phosphate pathway contribute to the defence response by providing precursors or cofactors for other biosynthetic pathways, e. g. erythrose-4-phosphate and pentose as precursors of lignin, amino acids, nucleic acids, and coenzymes potentially required under stress condition [20]. The G6PDH and 6PGDH by producing NADPH also support the ascorbate glutathione (Halliwel-Asada) cycle of chloroplasts and peroxisomes in the defence against oxidative stress incidence [11]. It was suggested, that these enzymes can be included in the group of catalase, superoxide dismutase, ascorbate peroxidase, and glutathione reductase/peroxidase [11]. In olive plants exposed to salt-induced oxidative stress NADPH-producing dehydrogenases (including G6PDH) were activated very probably in order to recycle NADPH necessary for the protection against oxidative damages [19]. In salt-stressed callus from *Carex moorcroftii* G6PDH played a central role in exclusion of $Na^+$ from cytoplasm, which is typical process for tolerant plants. The $Na^+$ exclusion represents the main mechanism for the adaptation of plants to salt stress [18]. G6PDH provides in this case NADPH for NADPH oxidase generating $H_2O_2$, which serves as a signal up-regulating plasma membrane $H^+$-ATPase activity and $Na^+/H^+$ antiporter level [18] (Figure 1). Under long-term salt stress G6PDH activity *via* its relationship with the plasma membrane NADPH oxidase is involved in both the reduced glutathione (GSH) maintenance and the $H_2O_2$ accumulation. The $H_2O_2$ again, as a signal molecule, increases the activities of G6PDH, glutathione reductase, and glutathione peroxidase, which finally enhance the glutathione cycle rate and result in the increased GSH levels. Thus G6PDH plays a critical role in maintaining cellular GSH levels under this type of stress [26]. The use of G6PDH-overexpressing cell lines clones and GSH depleting drug has shown that the G6PDH induction serves to maintain and regenerate the intracellular GSH pool [15]. The involvement of G6PDH in mediating NO production by nitrate reductase (requiring NADPH) necessary for establishing tolerance to salt was found in red kidney bean roots [16].

Activities of G6PDH and 6PGDH were also significantly increased in cold-hardened leaves of winter rye (*Secale cereale* L.). Contrary, activities of these enzymes decreased after the transfer of cold-hardened material from 4°C to 22°C.

An increased antioxidant content and high capacity for maintaining these antioxidants in a reduced state under the oxidative stress were found in cold hardened leaves that was imposed by the paraquat treatment. Therefore the oxidative pentose phosphate pathway may serve as a major non-photosynthetic source of NADPH production [22].

Hyperhydricity stress (previously referred to as vitrification) is characterized by reduced or retarded growth and morphological appearance indicating an excess of water content. This phenomenon in hyperhydric carnation leaves was accompanied by limited oxygen supply and by activation of G6PDH. Saher et al. suggested that adaptation of hyperhydric leaves of carnation to hypoxia stress consist in the induction of the oxidative pentose phosphate, fermentative pathways, and an activation of Halliwell-Asada cycle.

The production of NADPH in the reaction catalyzed by G6PDH provides not only the Halliwell-Asada cycle with NADPH, but also it substitutes a deficiency of the coenzyme in hyperhydricity tissues caused by decreased content of total chlorophylls and thereby decreased rate of photosynthesis. [23].

The G6PDH and 6PGDH activity was also increased in Cd-treated germinating pea seeds.

Explanatory notes relevant for this and following figures: Subcellular compartments are schematically indicated; key enzymes are shown in bold, underlined; NAD(P)H and ATP are highlighted; bold arrows indicate enzyme reactions, which can gain more importance under stress conditions; dashed arrows denote enzyme reactions that are in conditions of stress decreased; dotted arrows indicate possible metabolic flow; subsequent arrows denote multistep reactions and transporters are shown as black circles.

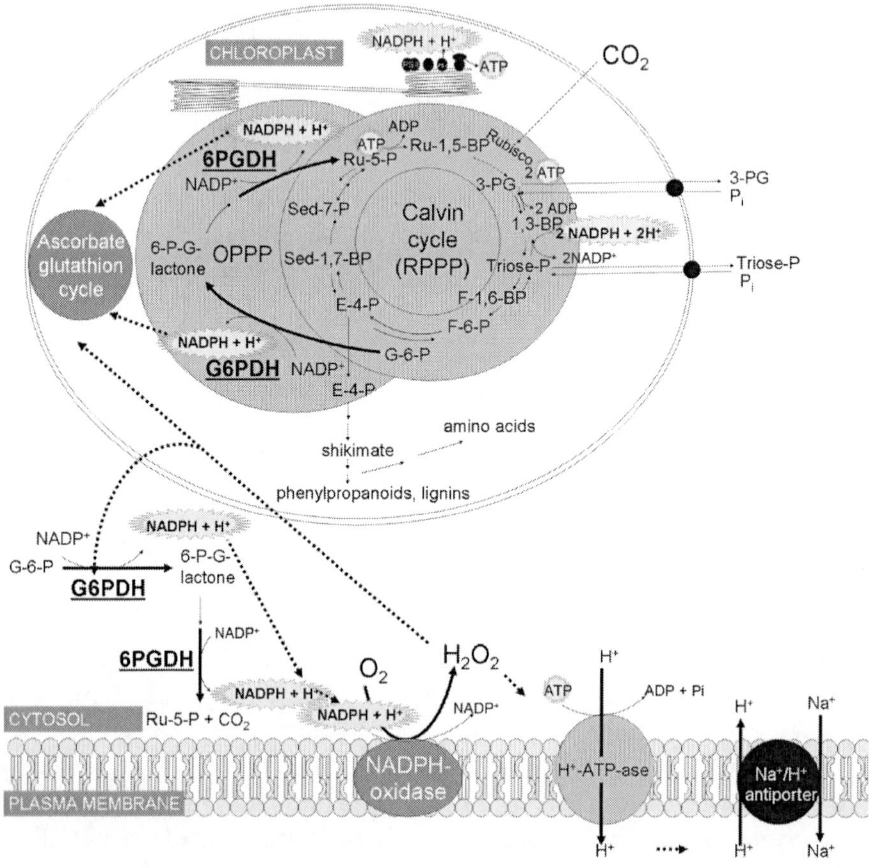


Figure 1. Simplified scheme of reactions catalyzed by G6PDH and 6PGDH and model illustrating hypothetical functions of G6PDH under abiotic stress conditions. Under salt stress, NADPH produced by G6PDH could be used for plasma membrane NADPH-oxidase generating $H_2O_2$. $H_2O_2$ could serves as a signal up-regulating plasma membrane $H^+$-ATPase activity and $Na^+/H^+$ antiporter level. Their up-regulation facilitate $Na^+$ efflux into apoplast and vacuoles thus decreased $Na^+$ toxicity to plants under saline stress [18]. Simultaneously, $H_2O_2$ as a signal increases activities of G6PDH and glutathione reductase, thus enhances ascorbate-glutathione cycle rate and increased GHS levels [26]. (1,3-BPG, 1,3-bisphosphoglycerate; E-4-P, D-erythrose-4-phosphate; F-6-P, D-fructose-6-phosphate; F-1,6-BP, D-fructose-1,6-bisphoshate; G-6-P, D-glucose-6-phosphate; 3-PG, 3-phosphoglycerate; 6-P-G-lactone, D-6-phosphoglucone lactone; OPPP, oxidative pentose phosphate pathway; RPPP, reductive pentose phosphate pathway; Ru-5-P, ribulose-5-phosphate; Ru-1,5-BP, ribulose-1,5-bisphosphate; Rubisco, ribulose-1,5-bisphosphate carboxylase/oxygenase; Sed-7-P, D-sedoheptulose-7-phosphate; Sed-1,7-BP, D-sedoheptulose-1,7-bisphosphate; Triose-P, triose-phosphate) Modified from [18] and [26].

As the heavy metal-induced respiratory damage results in a decrease of ATP production, G6PDH and 6PGDH are probably induced to compensate for the failure in energy supply similarly as during seed maturation [25]. An increase of the mRNA transcription and the activity of G6PDH was demonstrated in the case of aluminium stress in rye [24,27]. It is supposed that G6PDH supports the reduction of glutathione by glutathione reductase and by NADPH [24].

## 2.2. NADP-Isocitrate Dehydrogenase

NADP-isocitrate dehydrogenase (NADP-ICDH; EC 1.1.1.42) catalyzes the oxidative decarboxylation of isocitrate to form 2-oxoglutarate in the presence of $NADP^+$, simultaneously yielding NADPH and $CO_2$. The product, 2-oxoglutarate, providing the carbon skeleton that is required for the ammonium assimilation by means of glutamine synthetase and glutamate synthase, is a key metabolite in nitrogen and carbon metabolism [28,29] (Figure 2). Isocitrate dehydrogenase occurs in wide range of organism species. The presence of several isoenzymes located in different subcellular compartments, including the cytosol, chloroplast, mitochondria and peroxisome was shown [30]. The citric acid cycle enzyme NAD-ICDH (EC 1.1.1.41) was usually considered to be responsible for 2-oxoglutarate supply for amino acid synthesis. At present time NADP-ICDH is regarded as an alternative pathway in the case when large quantities of the 2-oxoglutarate are required [30]. A possible role for NADP-ICDH was suggested: the production of NADPH to support redox signalling or homeostasis in response to oxidative stress [31-34]. Recently it has been found that NADP-ICDH plays a role in the response to abiotic stress [19,35-41].

It is supposed that the function of NADP-ICDH is also very important under conditions of the pentose phosphate pathway impairment. For example, in pea nodules under water stress, when the activity of G6PDH and many other enzymes decreased, the NADP-ICDH was the only NADPH generating enzyme, whose activity increased [38]. The function of NADP-ICDH in addition to providing of carbon skeletons consists in the supply of NADPH and substrates for energy-yielding metabolism of bacteroids. It is considered, that at physiological conditions the later function NADP-ICDH fulfils together with NADP-dependent enzymes of the oxidative pentose phosphate pathway and malic enzyme. But under stress condition it seems that this function is provided only by NADP-ICDH, which simultaneously can compensate possible carbon and nitrogen imbalances in metabolism [38].

NADP-ICDH belongs to the top 40 salt stress up-regulated transcripts in chickpea roots [41].

The chilling stress affects not only enzymes involved in oxygen scavenging [42], but also enzymes involved in the ammonium assimilation and the carbon metabolism, namely NADP-ICDH [43]. An important role of NADP-ICDH in the production of carbon skeletons for the ammonium assimilation under stress condition was proved in rice roots exposed to the low temperature stress (from optimal 30°C to 20°C) [43]. This type of stress caused simultaneously an increase of NADP-ICDH and of glutamine synthetase (GS, EC 6.3.1.2) activity. In rice embryos, buds, and leaves NADP-ICDH activity was not inhibited by the chilling (5°C for one week) and during re-warming (25°C) experiments. After re-warming the NADP-ICDH activity increased in cultivar showed sensitivity to low temperatures as well as in low temperature tolerant cultivar. Results of this study indicate that NADP-ICDH together with glutathione reductase and G6PDH belongs to cold stable enzymes involved in antioxidant defence [42].

The increased citrate efflux was reported to enhance both the Al-resistance and P-acquisition efficiency in the transgenic lines [44,45]. There are two strategies to increase an internal citrate concentration followed by the enhanced citrate efflux: i) the increase of citrate synthase activity or ii) the decrease of NADP-ICDH activity. Even though Delhaize et al (2002) [45] achieved large reduction in NADP-ICDH activity by genetic manipulation of tobacco plants, they concluded that rather other factors can control citrate efflux, e.g. transport out of he roots.

Similarly, no major difference in the activity of NADP-ICDH was detected between Al-resistant and Al-sensitive line of triticale in root apices (triticale is synthetic hybrid between wheat and rye) [46]. On the other hand, the gene for NADP-ICDH was down-regulated in Al-resistant common bean genotype after 4 hours following the Al-treatment [47]. Higher expression of ICDH was observed in the sensitive cultivar of wheat than in the tolerant one after 6 hours of aluminium stress [48].

## 2.3. NADP-Malic Enzyme

NADP-malic enzyme (oxaloacetate decarboxylating) (NADP-ME, EC 1.1.1.40) catalyzes the oxidative decarboxylation of malate and $NADP^+$ as coenzyme to produce pyruvate, $CO_2$, and NADPH (Figure 2). Divalent metal ions ($Mg^{2+}$, $Mn^{2+}$) are required for this reaction [49,50]. NADP-malic enzymes are present in most of living organisms. Regulation of NADP-ME is depended on particular isoform and plant species; in tobacco it occurs *via* macroergic compounds such as ATP, GTP, ADP [51] and intermediates of citric acid cycle [52]. Two types of plant NADP-ME isoforms can be distinguished: the photosynthetic ones, which occur in CAM and some $C_4$ plants and play an important role in photosynthesis, and to the non-photosynthetic isoforms present in $C_3$ plants and non-photosynthetic tissues of $C_4$ plants. Genomes of the model $C_3$ plants *Arabidopsis thaliana* and *Oryza sativa* L. contain four *nadp-me* genes; three encoded cytosolic NADP-ME isoforms and one plastidic one [53,54].

Whereas in rice each NADP-ME gene has a different tissue specific and developmental profile, in *Arabidopsis thaliana* cytosolic NADP-ME2 and plastidic NADP-ME4 are constitutively expressed, while NADP-ME1 is restricted to roots and NADP-ME3 to trichomes and pollen. In *Nicotiana tabacum* L. plants two NADP-ME isoforms (one cytosolic and one plastidic) and the transcript for the third putative were also characterized. The all three transcripts are expressed in the mature tissue in a constitutive manner [52]. Non-photosynthetic isoforms are involved in the regulation of pH and in the connection with phosphoenolpyruvate carboxylase (PEPC, EC 4.1.1.31) and malate dehydrogenase (MDH, EC 1.1.1.37) in providing of important metabolites, particularly for the citric acid cycle and the nitrogen assimilation (Figure 2). One of the most important roles of non-photosynthetic NADP-ME is providing NADPH useful not only in biosyntheses but also in plant defence against stress [55-57].

Enhanced NADP-ME activity and/or increased amount of NADP-ME protein and/or transcripts were found in plants exposed to salt stress [19,52,54,58], drought [59]), ozone [60,61], under conditions of hyperhydricity [23], and heavy metal induced stress conditions [34]. Various types of stresses (polyethylene glycol, cold, darkness, salinity, abscisic acid, and salicylic acid increased NADP-ME activity in hexaploid wheat [62]. However, the transcription of gene for plastidic NADP-ME decreased in response to all stresses except to darkness and NaCl. Similarly, the gene for cytosolic NADP-ME was down regulated by all tested treatments.

Authors speculate that hexaploid wheat, which posses a large genome have more than two genes for NADP-ME [62]. In plants exposed to all these types of stress authors pointed out the role of NADP-ME in providing of NADPH. In the case of stress affecting water balance (drought, salt, and cold stress) NADPH serve for the biosynthesis of osmotically active compounds (e.g. proline, mannitol, glycine betaine). Moreover, under drought stress NADP-ME can modulate organic acid metabolism in guard cells and could control stomatal

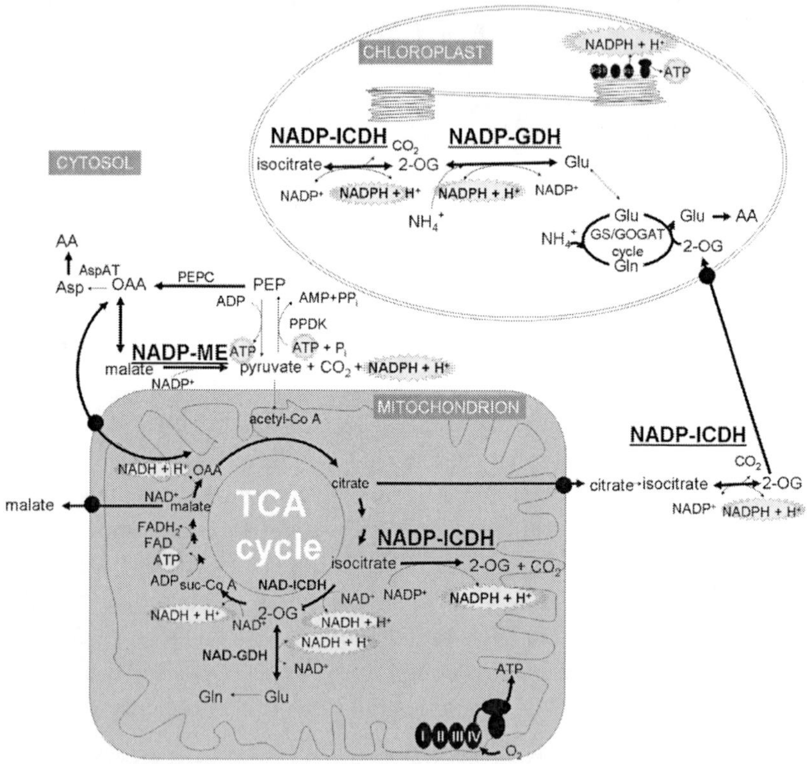


Figure 2. Scheme summarizing reactions catalyzed by NADP-ICDH, NADP-ME and NADP-GDH, their localizations, and functions. Under oxidative stress respiration and requirement of NADPH are increased. All above mentioned enzymes participate in stress affecting water balance (drought, cold, and salts) inducing biosynthesis of osmotically active compounds such as AA. NADP-ME could together with PEPC replenish citric acid cycle intermediates in the case of higher requirement of organic acids synthesis (e.g. phosphate and iron deficiency and aluminium toxicity). (AA, amino acids; Asp, L-aspartate; AspAT, aspartate aminotransferase; Glu, L-glutamate; Gln, L-glutamine; GS, glutamine synthetase; GOGAT, glutamate synthase; NAD-GDH; NAD-glutamate dehydrogenase, NAD-ICDH, NAD-isocitrate dehydrogenase; OAA, oxaloacetate; 2-OG, 2-oxoglutarate; PEP, phosphoenolpyruvate; PEPC, phosphoenolpyruvate carboxylase; PPDK, pyruvate, phosphate dikinase; suc-CoA, succinyl coenzyme A; TCA, tricarboxylic acid cycle). According to [55].

closure by degradation of malate [63]. In hyperhydric tissues exposed to conditions of limited oxygen, in which the fermentative metabolism is induced and pyruvate kinase activity is reduced, NADP-ME with PEPC can participate in reaction converting phosphoenolpyruvate to pyruvate [23] (Figure 2). Under conditions of cadmium stress the production of NADPH probably helps to establish the redox balance in the early-activated antioxidative defence mechanism [34].

## 2.4. NADP-Glutamate Dehydrogenase

Glutamate dehydrogenase (GDH) catalyzes a reversible reaction of the reductive amination of α-oxoglutarate to produce glutamate in the presence of cofactor NAD(P)H [64] . Whereas bacteria have only NADP-GDH, higher plants and microalgae contain both the NAD-GDH isoenzyme, which is a constitutive isoenzyme localized in mitochondria and the chloroplastic light-dependent, ammonium-inducible NADP-GDH (EC 1.4.1.3-4) [65] (Figure 2). Previously, GDH was considered as the main enzyme involved in ammonium assimilation. Since the discovery of the glutamine synthetase (GS, EC 6.3.1.2)/glutamate synthase (GOGAT, EC 1.4.1.14; EC 1.4.7.1) cycle a role of GDH in carbon metabolism rather than in nitrogen one was suggested. Physiological, molecular, and genetic studies have all shown that GS/GOGAT cycle function as the primary routes of ammonia assimilation in plants [64]. GS catalyzes the formation of glutamine from ammonium and glutamate. Subsequently, GOGAT mediates the transfer of the amido group of glutamine to α-oxoglutarate. One of the two molecules of glutamate that is formed represented yield of this cycle and can act as substrate for transamination enzymes while the other regenerate an ammonium acceptor during another round of the GS/GOGAT cycle (Figure 2) [66]. In addition GDH has a lower affinity for ammonium compared to GS [67]. Therefore it has been suggested that GDH is more active in the deamination rather than amination reaction in land plants [68], especially under conditions when ammonium ions are remobilized, for example during germination, senescence or carbon starvation [66]. Catabolism of glutamate by GDH could provide carbon skeletons for citric acid cycle as well as energy production during carbon and energy deficit. However, in some species function of GDH is related to a complementary role to GS/GOGAT in re-assimilation of ammonia excess during stress conditions or during a specific stage of the plant development [64]. GDH is supposed to play an important role in the control of ammonia assimilation in some species of algae [68].

For example, thermal stress (30°C) in a sterile mutant of marine alga (*Ulva pertusa*) caused an activation of NADP-GDH, which probably compensates the thermally induced decrease in nitrogen assimilation efficiency [68,69]. Several transgenic plants containing overexpressed bacterial [64] or fungal [66] NADP-GDH were prepared to modulate nitrogen metabolism. Transgenic tobacco plants containing the bacterial gene for NADP-GDH contained an increased amount of total carbohydrates than controls. Furthermore, these plants were more tolerant to herbicide (GS inhibitor); the introduced bacterial NADP-GDH partially substituted the reduced GS activity [64]. Transgenic tomato lines over expressing fungal NADP-GDH had increased NADP-GDH aminating activity, which was consistent with the increased levels of glutamate in fruits compared to non-transgenic controls [66]. Transgenic *Zea mays* plants overexpressing bacterial gene for NADP-GDH had improved drought tolerance and were also more tolerant to GS inhibitor. In addition, germination and grain biomass production in seasons with significant water deficit were increased in these plants [67]. Increased tolerance to water deficit by maintaining of water potential was found in transgenic *Nicotiana tabacum* L. plants containing bacterial gene for NADP-GDH. As under water stress conditions amino acids as osmolytes, osmoprotectants and scavengers of hydroxyl radicals are synthesized, NADP-GDH in these plants probably participates in enhancing of nitrogen availability and in providing carbon skeletons and reduction equivalents [70].

## 2.5. Non-Phosphorylating NADP-Glyceraldehyde-3-Phosphate Dehydrogenase

Non-phosphorylating glyceraldehyde-3-phosphate dehydrogenase (NP-GAPDH; EC 1.2.1.9) is a cytosolic enzyme catalyzing the irreversible oxidation of D-glyceraldehyde-3-phosphate into 3-phosphoglycerate coupled to $NADP^+$ reduction to NADPH [71] (Figure 3). NP-GAPDH is a member of aldehyde dehydrogenase superfamily (ALDHs). The Arabidopsis genome contains a single gene that encodes this enzyme [72]. NP-GAPDH plays a crucial role in a shuttle system for the indirect transport of photosynthetically generated NADPH from the chloroplast to the cytosol. In this process, glyceraldehyde-3-phosphate produced in the Calvin cycle is transported (*via* the chloroplast phosphate translocator) to the cytosol, where it is oxidized in the presence of $NADP^+$ to 3-phosphoglycerate by NP-GAPDH and NADPH is produced [73]. The reduced coenzyme generated in the cytosol may be used to meet several biosynthetic requirements [74]. This role is strictly related to photosynthetic metabolism in plant cell; however NP-GAPDH was also detected in plant non-photosynthetic tissues. In these cells NP-GAPDH could couple NADPH production needed for anabolic reactions with glycolysis [71]. In heterotrophic cells of wheat NP-GAPDH is post-translationally regulated by reversible phosphorylation that influences its enzyme activity and thus NP-GAPDH could regulate the NADPH production in the cytosol of these cells [71]. *Arabidopsis thaliana* plants lacking NP-GAPDH show a delay in growth, number differentially expressed genes and changed enzyme activities. Therefore this enzyme has a relevant role in plant growth and development [75].

NP-GAPDH participates in response to oxidative stress [73]. The oxidative condition imposed by methylviologen caused up to 2-fold increase in NP-GAPDH activity in maize and wheat seedling. There was found out in this study that the enzyme could be relatively protected against inactivation and thus it could be considered as a main producer of NADPH in cytoplasm of plant cells.

The activity of NP-GAPN was extremely low both, in the control and in the freeze-injured orange fruit (*Citrus sinensis*), suggesting a minor role of this enzyme in fruit metabolism [76].

Rice plants exposed to drought, submergence stress or the abscicis acid treatment showed an enhanced accumulation of NP-GAPDH transcripts [77]. The induction of expression of the genes involved in glycolysis and alcoholic fermentation (e.g. NP-GAPDH, enolase, alcohol dehydrogenase, and pyruvate decarboxylase) is essential for anaerobic tolerance under conditions of submergence stress. Results of this study showed that the enhanced rates of glycolysis are one of the immediate cellular responses to the water deficit; probably due to an importance of energy production during stress recovery [77]. Gao et al. (2000) have observed a significant increase in NP-GAPDH activity with increased levels of salt stress (50 mM and 150 mM) in premature and mature leaves of celery, but not in older leaves [74]. Simultaneously a close relation of NP-GAPDH and M6PR and their mRNA transcription levels in response to salt stress was determined. Therefore it is suggested that NP-GAPDH supplies NADPH for mannitol biosynthesis under tight metabolic control in plants synthesizing mannitol as transport metabolite [74] (Figure 3).

NP-GAPDH can be also involved in the phosphorus deficiency stress. In contrast to GAPDH involved in glycolysis, NP-GAPDH does not use $P_i$ as a substrate and directly produces 3-phosphoglycerate: by this way ATP generation reaction and the utilization of additional $P_i$ are bypassed [78] (Figure 3).

The function of NP-GAPDH and the presence of a pyrophosphate ($PP_i$)-dependent phosphofructokinase was suggested to be a result of adaptations of photosynthetic eukaryotes to the P-deficiency [78]. The extractable activity of NP-GAPDH in phosphate starved *Brassica nigra* suspension cells was 10-fold higher than in Pi-fed cells [79]. However in rice leaves grown under phosphorus deprivation an increase in the relative expression of NP-GAPDH was not observed.

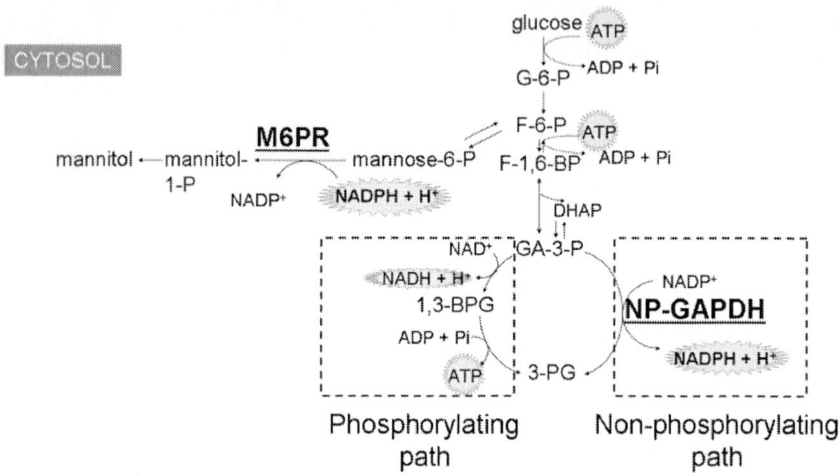


Figure 3. Scheme of the reaction catalyzed by NP-GAPDH, which is a part of non-phosphorylating alternative pathway of glycolysis, significant above all under phosphate deprivation and reaction catalyzed by M6PR directed to synthesis of mannitol, osmotically active compound. (1,3-BPG, 1,3-bisphosphoglycerate; DHAP, dihydroxyacetone phosphate; GA-3-P, D-glyceraldehyde-3-phosphate; glucose, D-glucose; G-6-P, D-glucose-6-phosphate; F-6-P, D-fructose-6-phosphate; F-1,6-BP, D-fructose-1,6-bisphosphate; mannitol-1-P, D-mannitol-1-phosphate; mannose-6-P, D-mannose-6-phosphate; 3-PG, 3-phosphoglycerate). According to [78].

On the contrary, NAD-GAPDH relative expression under the same conditions was significantly increased [80]. It means that the function of NP-GAPDH as a glycolytic bypass is not universal for all plants [80].

# 3. Enzymes Utilizing NADPH Under Abiotic Stress

## 3.1. Enzymes of Biosynthetic Pathways

### *3.1.1. Enzyme of Shikimate Pathway: Shikimate Dehydrogenase*

Shikimate dehydrogenase (SDH; EC 1.1.1.25), which catalyzes reversible conversion of dehydroshikimate and NADPH to shikimate and $NADP^+$, is present in bacteria, fungi, plants, and apicomplexa parasites, but is absent in animals and humans. In plants, this enzyme possesses also the 3-dehydroquinate dehydratase (DHD, EC 4.2.1.10) activity (abbreviated as DHD/SDH) [81]. While Arabidopsis plants have only a single DHD/SDH gene, tobacco plants possess two genes (plastidic and cytosolic) [82].

The Arabidopsis gene for DHD/SDH is important in the plant reproduction: it was shown to be required for the female gametophyte development and function [82,83]. The RNAi mediated reduction of expression of plastidic DHD/SDH in tobacco leads to severe phenotypic and metabolic alternation: 40% reduction of activity resulted in a strong decrease in growth parameters indicating an importance of the enzyme for plant development. Further reduction of DHD/SDH activity led to a lethal phenotype. Cytosolic tobacco DHD/SDH protein is capable of shikimate biosynthesis, but cannot supplement a loss of the plastidical pathway in the transgenic plants. Therefore it is likely that the role of cytosolic DHD/SDH *in vivo* is different from that of plastidic enzyme [81]. Plastidic DHD/SDH catalyzes the third and the fourth step of the shikimate pathway, which can be regarded as one of the main pronounced biosynthetic pathways of plants (Figure 4). This pathway is not restricted only to generation of amino acids for protein biosynthesis, but also it provides precursors for a large variety of other compounds formed by plants in large quantities, particularly phenylpropanoids (flavonoids and lignins) [84]. An importance of this pathway in plants is accented by the fact that about 20% of the carbon fixed in the photosynthesis flows through the shikimate pathway [85]. The shikimate pathway initiates from phosphoenolpyruvate (PEP) and erythrose-4-phosphate (E-4-P) and comprises seven reactions catalyzed by six enzymes (due to DHD/SDH as a bifunctional enzyme) to produce chorismate (Figure 4). Owing to its absence in humans, shikimate pathway is often target of herbicides such as glyphosate [86] and anti-tuberculosis drugs. Chorismate, the final product of the shikimate pathway, is the common precursor of the three aromatic amino acids phenylalanine, tryptophan, and tyrosine. These aromatic compounds represent central molecules in plant metabolism and they are essential substances in the diet of humans and monogastric livestock, which are not able to synthesize them [82,85]. Aromatic amino acids produced by shikimate pathway are utilized for the protein synthesis as well as they play pivotal role in the formation of phenylpropanoids (comprising simple phenols, flavonoids, stilbenes, tannins, lignans, and lignin) and also other compounds (e.g. phytoalexins, alkaloids, and glucosinolates) [84] (Figure 4). Plant secondary metabolites play an important role in the plant development as well as in the plant protection against various stresses.

Phenolic compounds play not only an important role in ROS removal, but also they act as metal chelators [87]. Among phenolic acids, protocatechuic acid is characterized by high chelating ability and chlorogenic acid protects against lipid peroxidation [88].

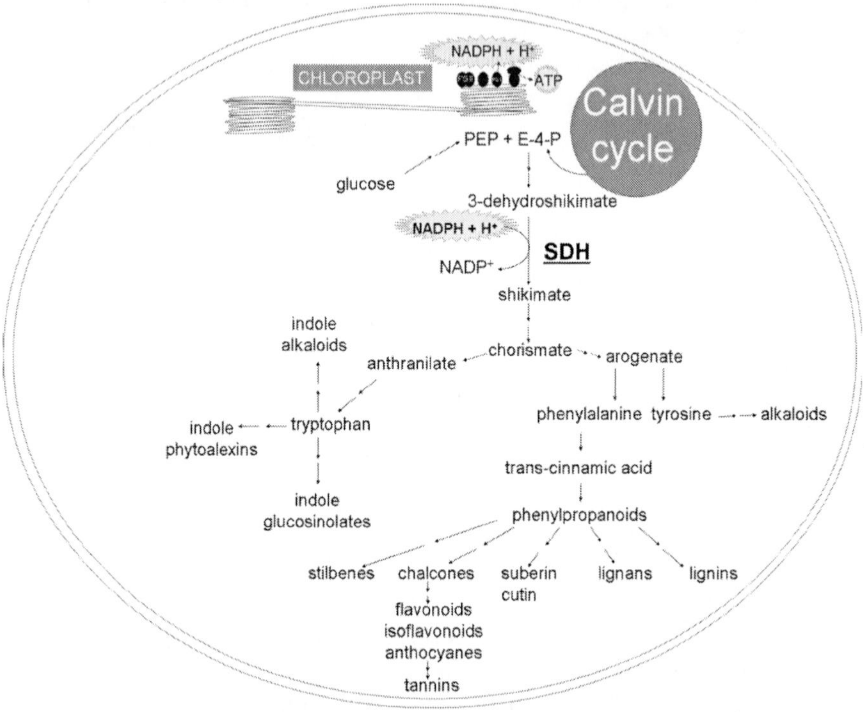


Figure 4. Simplified scheme of shikimate pathway, which plays a pivotal role in biosynthesis of aromatic amino acids (phenylalanine, tyrosine, and tryptophane), phenylpropanoids (stilbenes, chalcones, flavonoids, isoflavonoids, anthocyanes, tannins, suberins, cutins, lignans, and lignins) and secondary compounds (phytoalexins, alkaloids, and glucosinolates). (E-4-P, D-erythrose-4-phosphate; PEP, phosphoenolpyruvate). Modified from [1].

Pb stress increased the SHD activity in leaves of *Vallisneria natans* and also increased the content of total phenolic compounds and flavonoids [87]. SDH activity was also enhanced in Cu-exposed populations of *Silene vulgaris* [89] and in Ni-exposed *Matricaria chamomilla* plants [88].

The metabolism of phenolic compounds was substantially less affected by the Ni excess in comparison to the Cu one. However, a release of phenols decreases Cu uptake by complexation in the rhizosphere [88]. Pepper plants growing in a nutrient solution with an excess of Cu, showed an increase in the SDH activity in hypocotyl that is associated with the accumulation of soluble phenolics and lignins [90]. Different types of abiotic stresses such as mineral deficiency, drought, ultraviolet-B irradiation, mechanical injuries, and heavy metals as well as biotic stress result in changes of lignin content in several plants. An increased lignification seems to be defence response to stress in the case of many plants [91]. An increase of the SHD activity was also observed in the salicylic acid treated roots of *Panax ginseng* [92] and in strawberries fruit during storage at 1°C. In the later case the enhanced anthocyanin content was observed [93].

### *3.1.2. Enzymes Biosynthesizing Osmotically Active Compounds*

#### 3.1.2.1 Mannose-6-Phosphate Reductase

Biosynthesis of osmotically active compounds represents for many plants exposed to the water stress the possibility, how to decrease a water potential and to maintain a positive turgor pressure within the cell. The polyols (sugar alcohols such as sorbitol, mannitol, glycerol, and pinitol) and several related derivatives were the first class of compounds to be considered as compatible solutes, which play roles in the stress protection [94]. Since mannitol in many species also serves as transport saccharide, its biosynthesis is especially important. Photosynthetic partitioning of carbon between saccharose, sugar alcohols, and starch depends on the species, stage of the leaf development and environmental factors (e.g. salinity) [95]. The key enzyme of synthetic pathway is M6PR (EC, 1.1.1.224) requiring NADPH; this pathway starts with isomeration of fructose-6-phosphate to mannose-6-phosphate by mannose-6-phosphate isomerase, then it continues with the reduction of mannose-6-phosphate to mannitol-1-phosphate by M6PR and finally mannitol-1-phosphate is dephosphorylated of to mannitol [95] (Figure 3).

The salinity increased the activity of M6PR and the concentration of mannitol in the celery plant due to the shift in photosynthetic carbon partitioning (the concentration of sucrose and starch on the other hand decreased). A marked increase in the mannitol content, as an active adaptation to osmotic changes, was found in leaves and roots of olive plants subjected to water deficit [96], in coffee genotypes subjected to cold stress [97] and in leaves of celery plants stressed by excess macronutrients [98].

Recently it has been observed that the presence of M6PR gene in transgenic plants does not mean only the synthesis of mannitol as osmoprotectant. The presence of M6PR as transgene has also far-ranging effects on the expression of many redox-, stress- and cell wall related genes, which are up-regulated in the absence of stress [99]. The Arabidopsis plants transformed with M6PR gene from celery exhibited dramatically increased salt tolerance: the reduced salt injury and the suppressed inhibition of vegetative growth were observed. Based on the findings that relatively few additional genes were altered in the salt-stressed M6PR transgenic plants as compared to salt-stressed wild type plants, authors suggest that M6PR may participate in pre-adaptive changes facilitating the plant response to salt stress [99]. Therefore, M6PR was designated as a positive selection marker gene that selectively promotes the growth of the transformed cells from the non-transformed ones [100]. However, this phenomenon depends on plant species: in tobacco M6PR could be a negative marker inhibiting growth of transformed cells [100]. M6PR was also a target for genetic engineering in parasitic plant *Orobanche aegyptiaca*. Suppression of M6PR through the production of homologous dsRNA sequences in the transgenic host plant caused a significant decrease of mannitol level and a significant increase in the percentage of dead *O. aegyptiaca* tubercles on the transgenic host plant [101].

#### 3.1.2.2. $_{\Delta}{}^{1}$-Pyrroline-5-Carboxylate Synthetase, $_{\Delta}{}^{1}$-Pyrroline-5-Carboxylate Reductase

It is well known that plants accumulate under drought and salt stress amino acid proline [102-105]. The enhanced concentration of proline can be caused by its increased biosynthesis or decreased degradation (Figure 5). Proline synthesis starts from glutamate that is first phosphorylated to γ-glutamylphosphate and then reduced to glutamate semialdehyde; this intermediate is spontaneously cyclized to $_{\Delta}{}^{1}$-pyrroline-5-carboxylate. In plants these reactions are catalyzed by bifunctional enzyme ATP and NADPH dependent $_{\Delta}{}^{1}$-pyrroline-5-carboxylate synthetase (5PCS, EC 2.7.2.11/ 1.2.1.41), which consists of two domains, N-terminal with γ-glutamyl kinase activity and C-terminal with γ-glutamylsemialdehyde dehydrogenase activity. The resulting compound is then reduced to proline *via* $_{\Delta}{}^{1}$-pyrroline-5-carboxylate reductase (5PCR, EC 1.5.1.2), which is also dependent on NADPH [106] (Figure 5). Proline has a role not only as an osmolyticum, but it protects macromolecules especially proteins against denaturation, further it can act as a scavenger of the singlet oxygen and hydroxyl radical. Proline molecule seems to be important for plants in the stress recovery; it can serve as the source of carbon skeletons, of nitrogen and of reducing equivalents necessary for the biosynthesis of important metabolites or respiration [102,103].

The rate limiting enzyme in the proline biosynthesis is P5CS; in Arabidopsis plants this enzyme is encoded by two genes with non redundant functions. P5CS1 is involved in the proline accumulation related to stress conditions whereas P5CS2 responses to various developmental stages (e.g. flowering, seed development) [107,108].

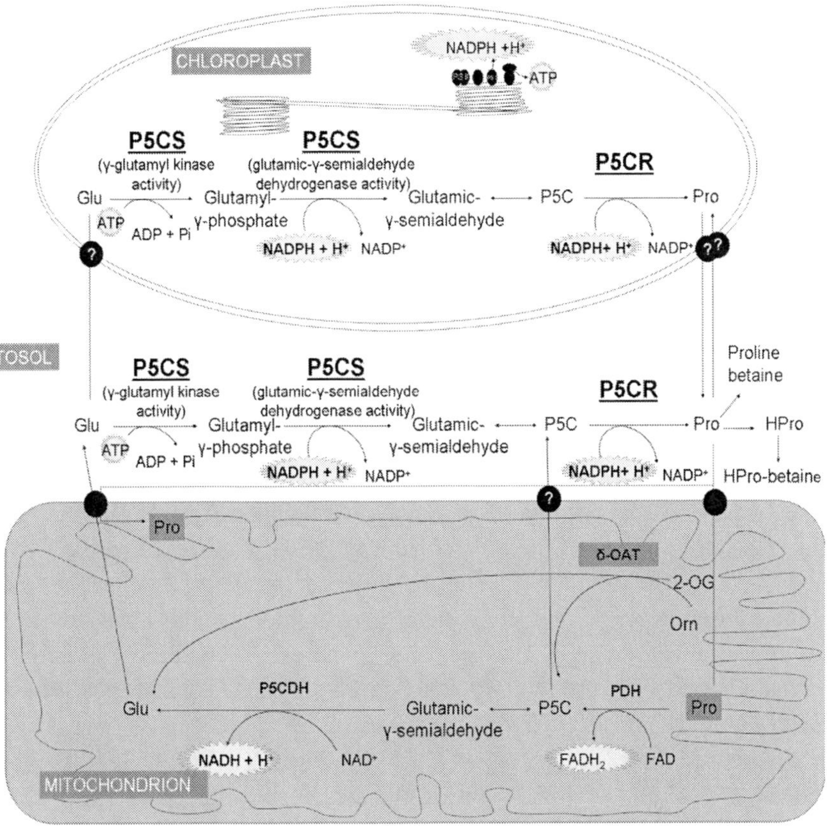


Figure 5. Scheme of involvement of P5CS and P5CR in proline metabolism and transport. Transporters are shown as black circles, proposed transporters are indicated by questionmarks. Biosynthesis of proline (Pro), hydroxyproline (HPro), proline betaine, and hydroxyproline betaine (HPro-betaine) is especially important under stress affecting water balance. (Glu, L-glutamate; 2-OG, 2-oxoglutarate; δ-OAT, ornithine-δ-aminotransferase; Orn, ornithine; PDH, proline dehydrogenase; P5C, $_{\Delta}{}^{1}$-pyrroline-5-carboxylate; P5CDH, 5PC dehydrogenase). According to [103] and [102].

The importance of the proline accumulation in plants grown under high salt conditions was proved in transgenic tobacco plants with overexpressed gene for P5CS. Transgenic plants grown under salt stress conditions synthesized 10 times more proline and were characterized by an enhanced biomass production and flower development [109]. Similar results were obtained also in other plant species [102].

### *3.1.3. Enzyme Participating in Biosynthesis of Isoprenoids: 3-Hydroxy-3-Methyl-Glutaryl-CoA Reductase*

Plants produce a wide variety of isoprenoid compounds. These compounds are involved in various biological processes, including the synthesis of membrane sterols, plant growth regulators, phytohormones, and some pharmaceutically important terpenoid compounds [93]. Furthermore, isoprenoids have been proved to play significant roles in plant-environment interaction including plant-plant, plant-insect, plant-microbe, and plant-animal communications as well as in plant-abiotic stress [110]. In plants, it is apparent that there are two distinct biosynthetic pathways leading to the synthesis of isoprenoid common precursor, isopentenyl diphosphate (IPP) and its isomer dimethylallyl diphosphate (DMAPP): (1) mevalonate (MVA) pathway localized in cytoplasm and (2) 2-C-methyl-D-erythritol-4-phosphate (MEP) pathway also called 1-deoxy-D-xylulose 5-phosphate (DXP) pathway, localized in chloroplasts (Figure 6) [93,111,112].

3-hydroxy-3-methylglutaryl-coenzyme A reductase (HMGR, EC 1.1.1.34) catalyzes the first step of the MVA pathway by converting 3-hydroxy-3-methyl-glutaryl-coenzyme A to MVA using NADPH as a cofactor (irreversible reaction) [113]. Sesquiterpens, triterpens, sterols, ubiquinone, and brassinosterols are biosynthesized *via* the MVA pathway, whereas gibberellins, abscisic acid, carotenoids and chlorophyll side chains are synthesized *via* the MEP pathway. However, the MVA and MEP pathways are not independent, as metabolic flow occurs between these pathways [110,112]. On the other hand, the direction predominates from the MEP pathway located in plastids to cytosol [110]. The physiological functions of plant HMGRs have been characterized using specific inhibitors such as lovastatin or mevinolin and tobacco cultured cells. HMGR was found out to be essential for biosynthesis of cytokinins (strong antagonists of senescence); an inhibition of HMGR reduced the cytokinin content and the cell proliferation [112].

Arabidopsis contains two differentially expressed HMGR genes that encode three HMGR isoforms, which exhibit distinct spatial and developmental patterns of expression and respond differentially to environmental stimuli. Previously, exclusive targeting of all three isoforms to ER was suggested. Recently it has been found that HMGR is predominantly localized also within spherical vesicular structures that are located in the cytoplasm and within the central vacuole [114]. HMGR activity is regulated at many levels, such as transcription, translation, regulated proteolytic degradation and kinase-mediated phosphorylation [115].

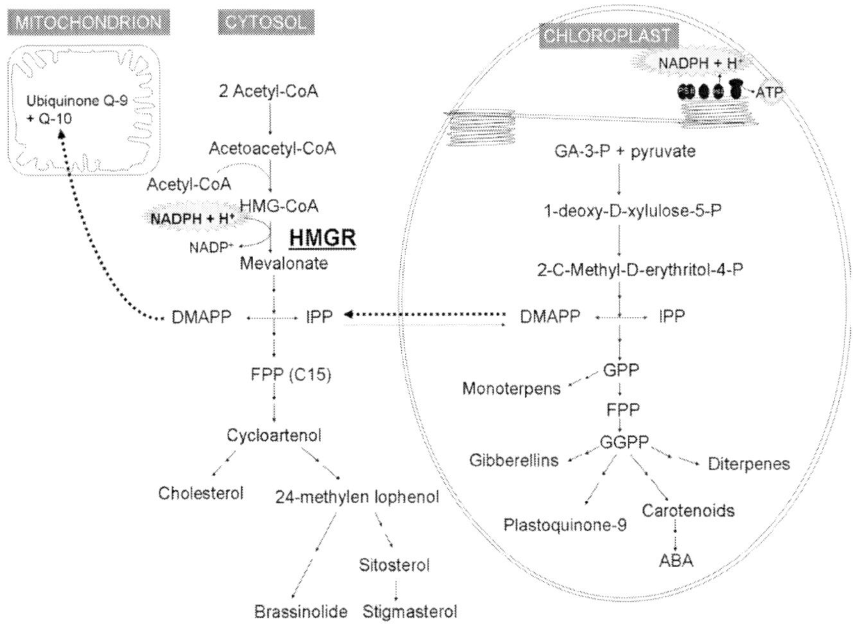


Figure 6. The role of HMGR in isoprenoid biosynthesis, simplified scheme. In cytosol occurs mevalonate (MVA) pathway and in plastids 2-C-Methyl-D-erythritol-4-P (MEP) pathway both leading to synthesis of isopentenyl diphosphate (IPP) as a precursor of isoprenoid biosynthesis. HMGR is a part of MVA pathway, which provides IPP for synthesis of sterols (sitosterol, stigmasterol, and cholesterol) and brassinolide in cytosol and ubiquinone structures in mitochondria. MEP pathway directs to synthesis of monoterpens, diterpens, gibberellins, carotenoids, abscisic acid, and plastoquinone-9. Arrows between cytosolic and plastidic compartment represent metabolic flow between them (thicker arrow for more significant flux). (ABA, abscisic acid; 2-C-Methyl-D-erythritol-4-P, 2-C-Methyl-D-erythritol-4-phosphate; DMAPP, dimethylallyl diphosphate; 1-deoxy-D-xylulose-5-P, 1-deoxy-D-xylulose-5-phoshate; FPP, farnesyl diphosphate; GA-3-P, D-glyceraldehyde-3-phosphate; GPP, geranyl diphosphate; GGPP, geranylgeranyl diphosphate; HMG-CoA, 3-hydroxy-3-methylglutaryl-CoA). According to [110].

Over-expression of the gene for HMGR in transgenic plants increased the amount of the end product sterols, but did not affect the accumulation of carotenoids and chlorophylls [113]. Enhanced expression of HMGR in *Levandula latifolia* increased content of stigmasterol and β-sitosterol [113], these compounds posses according to several studies a high antioxidant activity [110]. Thus, an ability of plants to accumulate sitosterol would be applicable for crops to resist stresses caused by $H_2O_2$ (e.g. ultraviolet-B light) [110]. Loss of activity of HMGR in Arabidopsis mutant results in dwarfing, reduced sterols levels and thus to early senescence [112]. The HMGR activity in potato tuber was increased by cutting [111].

## 3.2. Enzymes Involved in Antioxidant System

Plants possess multiple defence mechanisms involving small molecular antioxidants (such as ascorbate, glutathione, and tocopherol) and enzymes, which prevent the formation of reactive oxygen species or participate in their removal. ROS are frequently produced in plants exposed to abiotic stress [116,117]. The main mechanism for the detoxification of ROS in plants is the ascorbate-glutathione cycle, also known as the Halliwell-Asada pathway, involving four enzymes ascorbate peroxidase (APOD, EC 1.11.1.11), monodehydroascorbate reductase (MDHAR, EC 1.6.5.4), dehydroascorbate reductase (DHAR, EC 1.8.5.1) and glutathione reductase (GR, EC 1.8.1.7) [117-119]. Two of these enzymes (MDHAR and GR) are the NAD(P)H dependent enzymes (Figure 7). Most of the ascorbate-glutathione cycle enzyme activities are located in multiple organelles: chloroplasts, cytosol, mitochondria, and peroxisomes [116].

### *3.2.2. Monodehydroascorbate Reductase*

MDHAR, a FAD enzyme, catalyzes the regeneration of reduced ascorbate (monodehydroascorbate) to ascorbate using NAD(P)H as electron donor (Figure 7). Both NADH and NADPH serve as electron donors, with NADH being the preferred one [121]. MDHAR can use also alternative electron acceptors, e.g. ferricyanide, cytochromes, and phenoxyl radicals [122,123]. The fact that this enzyme makes use an organic radical as a substrate is unique [123]. In plants, the MDHAR activity has been described in several compartments, such as chloroplasts, cytosol, mitochondria, glyoxysomes and leaf peroxisomes.

*Arabidopsis thaliana* has five genes of MDHAR and one of them has multiple transcription starts that cause a dual targeting to chloroplast and mitochondria [123]. It is suggested that expression levels of MDHAR and dehydroascorbate reductase (DHAR) genes are regulated by the low ascorbate content and the light abundance, which affects the capacity of ascorbate recycling [124].

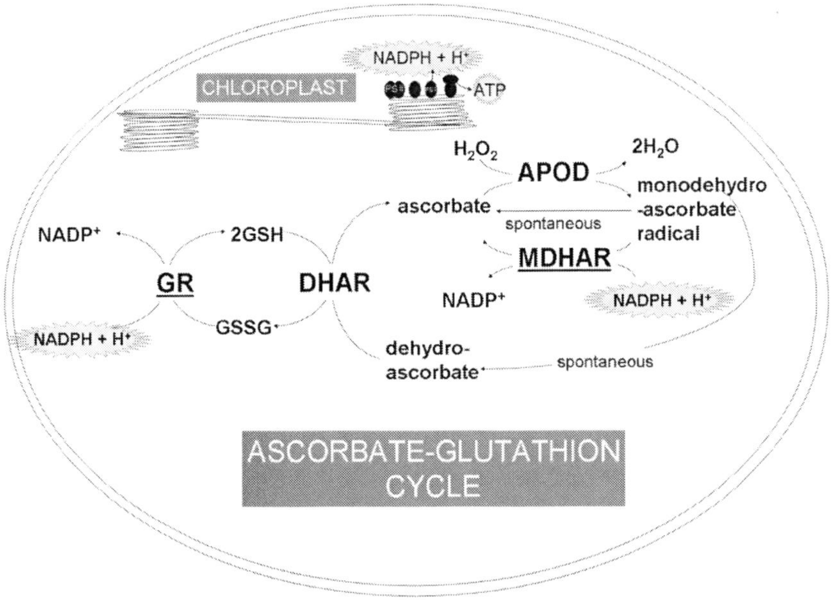


Figure 7. Ascorbate-glutathione cycle requiring the consumption of reduction equivalents NAD(P)H protects plant cells against damaging effect of reactive $H_2O_2$. Ascorbate-glutathione cycle starts with reduction of $H_2O_2$ by the enzyme ascorbate peroxidase (APOD), using ascorbate as electron donor and generating monodehydroascorbate free radical. Monodehydroascorbate can be directly reduced back to ascorbate by the enzyme MDHAR with NAD(P)H as electron donor. Alternatively, two molecules of monodehydroascorbate are spontaneously dismutated to ascorbate and dehydroascorbate. The latter is then reduced back to ascorbate using reduced glutathione, which, in turn is regenerated by glutathione reductase (GR) using electrons provided by NADPH [120]. DHAR, dehydroascorbate reductase; GSH, reduced glutathione; GSSG oxidized glutathione. Modified from [1].

Ascorbate is a major antioxidant with the ability to scavenge free radicals directly. It serves as important redox buffer in plants, as a cofactor of many enzymes and as a regulator of the cell division and growth and finally as a molecule for signal transduction in plants [125]. However it is considered that the primary function of ascorbate is to provide electrons for the reduction of $H_2O_2$ *via* APOD. By this reaction it protects the hydrogen peroxide sensitive enzymes such as enzymes of the Calvin cycle in the stroma of chloroplasts [125]. MDHAR also participates in the mediation of the photoreduction of dioxygen to superoxide radicals in the case when the enzyme substrate (monodehydroascorbate) is absent [123]. The regulation of ascorbate levels in cells is tightly controlled by the level of synthesis, recycling, degradation and transport of this molecule within the cell or between organs. With the MDHAR ability to regenerate directly ascorbate, this enzyme plays an important role in maintaining the reduced pool of ascorbate and serves as a vital component of the defence response to oxidative stress [124-126].

Transgenic tobacco plants overexpressing MDHAR posses higher MDHAR activity, increased ascorbate levels and enhanced tolerance to ozone, salt, and polyethylene glycol induced stress. Eltayeb et al. assumed that the enhanced tolerance of these transgenic plants in term of higher net photosynthesis could be explained by higher levels of ascorbate that protect and restore the photochemical activity of photosystem II and ascorbate maintain the redox status [125]. Transgenic tomato lines overexpressing chloroplastic MDHAR posses enhanced tolerance to the temperature and the methyl-viologen oxidative stresses [127]. Under these conditions ascorbate regenerated in chloroplast is needed not only for ascorbate peroxidase but also to protect photosynthetic apparatus by donating electrons to photosystem II, when a donation of electrons from water is disturbed as a result of heat stress [127]. In pea plants MDHAR activity in peroxisomes was up-regulated under conditions of high light intensity and cadmium stress. Peroxisomal MDHAR could protect cell against $H_2O_2$ leaking from peroxisomes and could be useful in reoxidation of endogenous NADH, which ensure a constant supply of $NAD^+$ for peroxisomal metabolism [123].

A role for MDHAR in the maintenance of ascorbate levels in tomato fruit under chilling stress conditions has been shown [126]. Oxidative stress affects MDHAR in these plants also at transcriptional level; an involvement of the enzyme probably consists in the detoxification of increased levels of ROS. In *Brassica campestris* plants the MDHAR mRNA was up-regulated in response to oxidative stress invoked by $H_2O_2$, salicylic acid, paraquat, and ozone [121]. Two cytosolic isoforms of MDHAR displayed significant up-regulation of mRNA levels in the sensitive genotype of wheat, whereas only one of them did in the tolerant cultivar [128].

Enhanced MDHAR activity has been reported in stressed plants many times. For the illustration, there are some examples published recently: in the leaves of *Malpighia glabra* plants under cold and salt stress condition [124], in leaves of *Brassica juncea* L. plants exposed to cadmium [129], in leaves of tomato (*Solanum lycopersicum* L. cv. Zarina) plants exposed to moderate drought [130], in *Eupatorium adenophorum* plants sensitive to high temperatures in conditions of low temperatures [131] and in phtalic acid stressed roots of *Malus prunifolia* [132].

### *3.2.1. Glutathione Reductase*

Glutathione reductase uses reductive power of NADPH to regenerate reduced glutathione (GSH) from oxidized glutathione (GSSG) (Figure 7). In fact, GR reduces the disulfide bond that links together two molecules of reduced glutathione (tripeptide $\gamma$-L-glutamylcysteinylglycine). GR is a flavoprotein, which has been found in all organisms investigated [118]. The Arabidopsis genome contains two GR genes. One gene codes for an organeller isoform, which is dual-targeted to chloroplasts and mitochondria. This isoform is essential for plant development. The second gene codes for the cytosolic isoform that seems to be responsive to several stress factors [3]. Three genes encoding GR have been described in *Oryza sativa* L.; a cytosolic and two chloroplastic [119]. The physiological role of GR in the cells is obviously related to the function of glutathione. Glutathione is an abundant metabolite in plants that has diverse and important functions, including control of cellular redox homeostasis, signal transduction, regulation of protein synthesis, sulfur assimilation and metabolism and detoxification of xenobiotics [3,133-135]. Many studies have demonstrated that GR plays an important role in protecting plants against oxidative stress, mainly by maintaining high GSH/GSSG ratio; increased GR activity was found e.g. in wheat roots exposed to Cd stress [133], in cotton leaves under NaCl stress [136] and in *Arabidopsis thaliana* leaves exposed to ozone stress [137]. The protection of plants against both the high and the low temperatures stresses consists in the prevention of the oxidation of enzymes and membranes by GR [138].

Drought stress induced an up-regulation of the expression of the cytosolic GR gene directly related to the intensity of the stress in both drought sensitive and drought tolerant *Vigna unguiculata* cultivar [117]. Oxidative stress induced by cadmium up-regulated GR activity by the induction of distinctive isoforms in wheat roots. An alternation in the pattern of GR isoenzymes were observed in pea roots subjected to cold, ozone, and greening treatment [139].

## 3.3. Regulatory Enzymes

### *3.3.1. Thioredoxin Reductases*

Thioredoxin systems are ubiquitous redox regulators that facilitate the reduction of a wide scale of other proteins *via* disulfide-exchange reactions [5]. Plants contain more types of thioredoxin systems than any other organisms [140]. First, the ferredoxin-thioredoxin system (FTS) of oxygenic photosynthesis located in chloroplasts, is composed of ferredoxin (Fdx), ferredoxin-thioredoxin reductase

(EC 1.8.7.2, FTR), and thioredoxins (Trx) f, m, x, and y [141]. This system is not NADPH-dependent but is involved in light/dark regulation of chloroplast enzymes and in oxidative stress responses [140]. The disulfide bonds in target enzymes are reduced using photons and electrons (from water) according to the sequence Light → Fdx → FTR → Trx (f, m, x, y) → Target Protein [141, 142].

Second, NADP-thioredoxin system (NTS) is localized in cytosol and mitochondria. NTS contains NADPH, NADP-thioredoxin reductase (EC 1.8.1.9, NTR), and thioredoxins. The enzyme catalyzes the transfer of electrons from NADPH to Trx h (in cytoplasm) or Trx h and o in mitochondria, according to the sequence NADPH → NTR → Trx h (o) → Target Protein. In Arabidopsis, NADPH-thioredoxin reductase A (NTRA) and NADPH-thioredoxin reductase B (NTRB) participates in the reduction of mitochondrial and cytosolic Trxs [3] (Figure 8). A system using NADPH for reductive reactions is also present in plant chloroplasts. This reaction is catalyzed by NADPH thioredoxin reductase C (NTR, designated NTRC), which act s as a bifunctional enzyme with N-terminal thioredoxin reductase domain and C-terminal Trx domain [141]. NTRC is present only in organisms with oxygenic photosynthesis. Due to a high affinity of NTRC to NADPH, this system might participate as an alternative electron donor for the chloroplast detoxification system especially during darkness. [143,144] (Figure 8).

The function of NTRC was proved in NTRC deficient plants. These transgenic Arabidopsis plants had some fenotypic alternations such as retarded growth, irregular cell morphology, chloroplast structure, and lower rate of photosynthesis, which were more pronounced, when the plants were grown under the short-day conditions [144].

Many thioredoxin target proteins were identified; some of them are involved in plant response to stress. Very important target enzymes are thioredoxin peroxidases (EC 1.11.1.15, peroxiredoxins). These enzymes are able to reduce $H_2O_2$ using a thiol group and they are reactivated by thioredoxins. Presence of peroxiredoxins is necessary for the protection of cell structures under optimal growth conditions but it is more pronounced under stress [145]. Another target protein is NADPH oxidase localized in mitochondrial membrane; this enzyme is activated in its reduced form. Beside enzymes of Calvin cycle, which are activated by FTR, also cytosolic NADP-ICDH and NADP-malate dehydrogenase are activated by NTR [141,146]. NADP-ICDH is involved in many types of stress (see section 2.2).

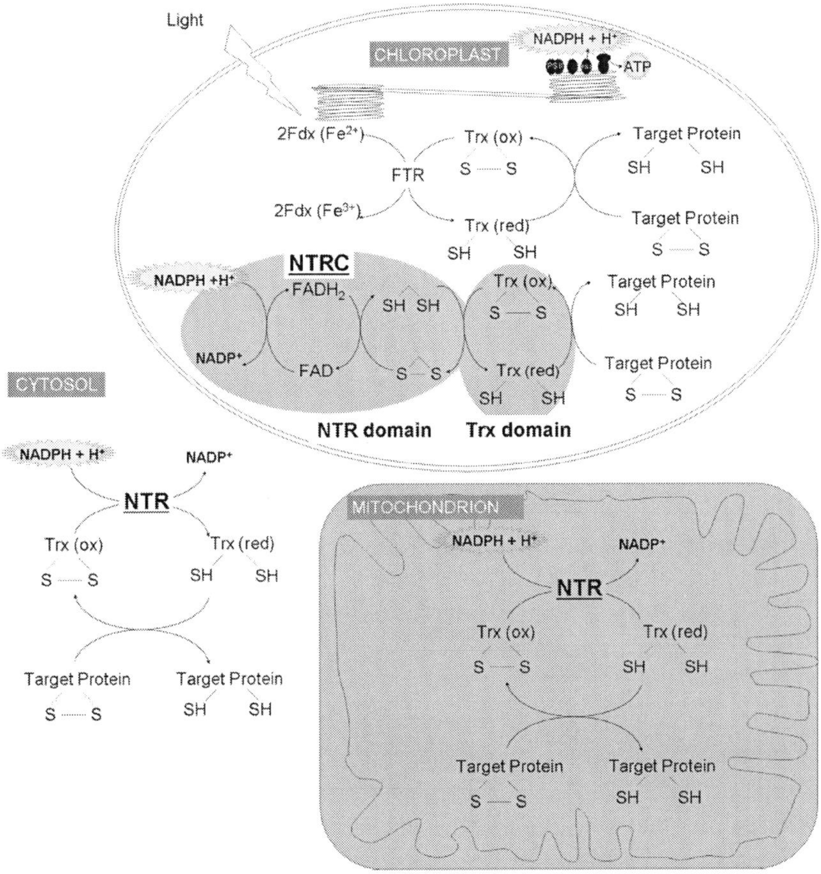


Figure 8. The role of NADP-thioredoxin reductase (NTR) and ferredoxin-thioredoxin reductase (FTR) in thiredoxin (Trx) reduction. In turn reduced Trx using reduction of disulfide bonds activated/deactivated a lot of various target proteins. In chloroplasts, Trx reduction occurs *via* FTR and depends on light availability due to reduced ferredoxin (Fdx ($Fe^{2+}$)). In mitochondria and cytosol, thioredoxins could be reduced by NTR. Specific bifunctional NTRC was found in chloroplast, which is formed by an NTR domain at N-terminus and a Trx domain at the C-terminus. Modified from [144,147].

# CONCLUSIONS

For plants under stress conditions it is very important to maintain the homeostasis within the cell. Plants have a lot of systems to regulate the reduction oxidation balance; reactions catalyzed by NADP-dependent enzymes described in this chapter participate in maintenance of constant NADP/NADPH ratio. In stress conditions the demand of NADPH could be enhanced due to biosynthesis of osmoprotectants, secondary metabolites, and of enhanced requirement for cofactors of antioxidant enzymes. The main metabolic pathways in individual plant species are identical, but the variability exists, especially in partitioning metabolites among the relevant pathways. The preferential utilization of a particular NADP dependent enzyme could be different in individual plant species. Therefore the effect of a stress type on a plant species (cultivars) needs a detailed enzymological study, which could help to select crop suitable for growing in non-optimal condition (e.g. high temperature, low water availability).

## REFERENCES

[1] Buchanan, B. B.; Gruissem., W. and Russel., J, R. (2000). *Biochemistry and Molecular Biology of Plants*. Rockville, Maryland: American society of Plant Physiologists.

[2] Pollak, N.; Dolle, C. and Ziegler, M. (2007). The power to reduce: pyridine nucleotides-small molecules with a multitude of functions. *Biochem. J.*, *402*, 205-218.

[3] Marty, L.; Siala, W.; Schwarzlander, M.; Fricker, M. D.; Wirtz, M.; Sweetlove, L. J.; Meyer, Y.; Meyer, A. J.; Reichheld, J. P. and Hell, R. (2009). The NADPH-dependent thioredoxin system constitutes a functional backup for cytosolic glutathione reductase in Arabidopsis. *Proc. Natl. Acad. Sci. USA*, *106*, 9109-9114.

[4] Noctor, G. (2006). Metabolic signalling in defence and stress: the central roles of soluble redox couples. *Plant Cell Environ.*, *29*, 409-425.

[5] Kirkensgaard, K. G.; Hagglund, P.; Finnie, C.; Svensson, B. and Henriksen, A. (2009). Structure of *Hordeum vulgare* NADPH-dependent thioredoxin reductase 2. Unwinding the reaction mechanism. *Acta Crystallogr. D Biol. Crystallogr.*, *65*, 932-941.

[6] Hayano-Kanashiro, C.; Calderon-Vazquez, C.; Ibarra-Laclette, E.; Herrera-Estrella, L. and Simpson, J. (2009). Analysis of gene expression and physiological responses in three mexican maize landraces under drought stress and recovery irrigation. *PLoS One*, *4*, 1-19.

[7] Atkin, O. K. and Macherel, D. (2009). The crucial role of plant mitochondria in orchestrating drought tolerance. *Ann. Bot.*, *103*, 581-597.

[8] Soares-Cordeiro, A. S.; Driscoll, S. P.; Arrabaca, M. C. and Foyer, C. H. (2010). Dorsoventral variations in dark chilling effects on photosynthesis

and stomatal function in *Paspalum dilatatum* leaves. *J. Exp. Bot.*, *62*, 687-699.

[9] Sade, N.; Gebretsadik, M.; Seligmann, R.; Schwartz, A.; Wallach, R. and Moshelion, M. (2010). The role of tobacco aquaporin1 in improving water use efficiency, hydraulic conductivity, and yield production under salt stress. *Plant Physiol.*, *152*, 245-254.

[10] Valenti, V.; Stanghellini, M. A. and Pupillo, P. (1984). Glucose 6-phosphate dehydrogenase isozymes of maize leaves: Some comparative properties. *Plant Physiol.*, *75*, 521-526.

[11] Corpas, F. J.; Barroso, J. B.; Sandalio, L. M.; Distefano, S.; Palma, J. M.; Lupianez, J. A. and Del Rio, L. A. (1998). A dehydrogenase-mediated recycling system of NADPH in plant peroxisomes. *Biochem. J.*, *330*, 777-784.

[12] Huang, J.; Zhang, H.; Wang, J. and Yang, J. (2003). Molecular cloning and characterization of rice 6-phosphogluconate dehydrogenase gene that is up-regulated by salt stress. *Mol. Biol. Rep.*, *30*, 223-227.

[13] Ishida, S. and Ashihara, H. (1993). Isoenzymes of phosphogluconate dehydrogenase (decarboxylating) from suspension-cultured *Catharanthus roseus* cells. *Phytochemistry*, *33*, 1307-1311.

[14] Scharte, J.; Schon, H.; Tjaden, Z.; Weis, E. and von Schaewen, A. (2009). Isoenzyme replacement of glucose-6-phosphate dehydrogenase in the cytosol improves stress tolerance in plants. *Proc. Natl. Acad. Sci. USA*, *106*, 8061-8066.

[15] Salvemini, F.; Franze, A.; Iervolino, A.; Filosa, S.; Salzano, S. and Ursini, M. V. (1999). Enhanced glutathione levels and oxidoresistance mediated by increased glucose-6-phosphate dehydrogenase expression. *J. Biol. Chem.*, *274*, 2750-2757.

[16] Liu, Y.; Wu, R.; Wan, Q.; Xie, G. and Bi, Y. (2007). Glucose-6-phosphate dehydrogenase plays a pivotal role in nitric oxide-involved defense against oxidative stress under salt stress in red kidney bean roots. *Plant Cell Physiol.*, *48*, 511-522.

[17] Pandolfi, P. P.; Sonati, F.; Rivi, R.; Mason, P.; Grosveld, F. and Luzzatto, L. (1995). Targeted disruption of the housekeeping gene encoding glucose-6-phosphate-dehydrogenase (*G6PD*) - G6PD is dispensable for pentose synthesis but essential for defense against oxidative stress. *EMBO Journal*, *14*, 5209-5215.

[18] Li, J.; Chen, G.; Wang, X.; Zhang, Y.; Jia, H. and Bi, Y. (2011). Glucose-6-phosphate dehydrogenase-dependent hydrogen peroxide production is involved in the regulation of plasma membrane $H^+$-ATPase and $Na^+/H^+$

antiporter protein in salt-stressed callus from *Carex moorcroftii*. *Physiol. Plant.*, *141*, 239-250.

[19] Valderrama, R.; Corpas, F. J.; Carreras, A.; Gomez-Rodriguez, M. V.; Chaki, M.; Pedrajas, J. R.; Fernandez-Ocana, A.; Del Rio, L. A. and Barroso, J. B. (2006). The dehydrogenase-mediated recycling of NADPH is a key antioxidant system against salt-induced oxidative stress in olive plants. *Plant Cell Environ.*, *29*, 1449-1459.

[20] Nemoto, Y. and Sasakuma, T. (2000). Specific expression of glucose-6-phosphate dehydrogenase (G6PDH) gene by salt stress in wheat (*Triticum aestivum* L.). *Plant Sci.*, *158*, 53-60.

[21] Wang, X.; Ma, Y.; Huang, C.; Wan, Q.; Li, N. and Bi, Y. (2008). Glucose-6-phosphate dehydrogenase plays a central role in modulating reduced glutathione levels in reed callus under salt stress. *Planta*, *227*, 611-623.

[22] Crecelius, F.; Streb, P. and Feierabend, J. (2003). Malate metabolism and reactions of oxidoreduction in cold-hardened winter rye (*Secale cereale* l.) leaves. *J. Exp. Bot.*, *54*, 1075-1083.

[23] Saher, S.; Fernandez-Garcia, N.; Piqueras, A.; Hellin, E. and Olmos, E. (2005). Reducing properties, energy efficiency and carbohydrate metabolism in hyperhydric and normal carnation shoots cultured *in vitro*: A hypoxia stress? *Plant Physiol. Biochem.*, *43*, 573-582.

[24] Milla, M. A.; Butler, E.; Huete, A. R.; Wilson, C. F.; Anderson, O. and Gustafson, J. P. (2002). Expressed sequence tag-based gene expression analysis under aluminum stress in rye. *Plant Physiol.*, *130*, 1706-1716.

[25] Smiri, M.; Chaoui, A. and El Ferjani, E. (2009). Respiratory metabolism in the embryonic axis of germinating pea seed exposed to cadmium. *J. Plant Physiol.*, *166*, 259-269.

[26] Wang, X.; Ma, Y.; Huang, C.; Li, J.; Wan, Q. and Bi, Y. (2008). Involvement of glucose-6-phosphate dehydrogenase in reduced glutathione maintenance and hydrogen peroxide signal under salt stress. *Plant Signal Behav.*, *3*, 394-395.

[27] Slaski, J. J. (1994). Differences in the metabolic responses of root-tips of wheat and rye to aluminum stress. *Plant and Soil*, *167*, 165-171.

[28] Yoon, J. J.; Hattori, T. and Shimada, M. (2003). Purification and characterization of NADP-linked isocitrate dehydrogenase from the copper-tolerant wood-rotting basidiomycete *Fomitopsis palustris*. *Biosci. Biotechnol. Biochem.*, *67*, 114-120.

[29] Alvarado, A. and Flores, M. E. (2003). Characterization and regulation of $NADP^+$-isocitrate dehydrogenase from *Saccharopolyspora erythraea*. *Biotechnol. Lett.*, *25*, 1175-1178.

[30] Palomo, J.; Gallardo, F.; Suarez, M. F. and Canovas, F. M. (1998). Purification and characterization of $NADP^+$-linked isocitrate dehydrogenase from Scots pine. Evidence for different physiological roles of the enzyme in primary development. *Plant Physiol.*, *118*, 617-626.

[31] Moller, I. M. and Rasmusson, A. G. (1998). The role of NADP in the mitochondrial matrix. *Trends Plant Sci.*, *3*, 21-27.

[32] Hodges, M.; Flesch, V.; Galvez, S. and Bismuth, E. (2003). Higher plant $NADP^+$-dependent isocitrate dehydrogenases, ammonium assimilation and NADPH production. *Plant Physiol. Biochem.*, *41*, 577-585.

[33] Marino, D.; Gonzalez, E. M.; Frendo, P.; Puppo, A. and Arrese-Igor, C. (2007). NADPH recycling systems in oxidative stressed pea nodules: A key role for the $NADP^+$ -dependent isocitrate dehydrogenase. *Planta*, *225*, 413-421.

[34] Smeets, K.; Cuypers, A.; Lambrechts, A.; Semane, B.; Hoet, P.; Van Laere, A. and Vangronsveld, J. (2005). Induction of oxidative stress and antioxidative mechanisms in *Phaseolus vulgaris* after Cd application. *Plant Physiol. Biochem.*, *43*, 437-444.

[35] Leterrier, M.; Del Rio, L. A. and Corpas, F. J. (2007). Cytosolic NADP-isocitrate dehydrogenase of pea plants: genomic clone characterization and functional analysis under abiotic stress conditions. *Free Radic. Res.*, *41*, 191-199.

[36] Lee, D. G.; Ahsan, N.; Lee, S. H.; Lee, J. J.; Bahk, J. D.; Kang, K. Y. and Lee, B. H. (2009). Chilling stress-induced proteomic changes in rice roots. *J. Plant Physiol.*, *166*, 1-11.

[37] Chaffei, C.; Suzuki, A.; Masclaux-Daubresse, C.; Ghorbel, M. H. and Gouia, H. (2006). Implication of glutamate, isocitrate and malate deshydrogenases in nitrogen assimilation in the cadmium-stressed tomato. *C. R. Biol.*, *329*, 790-803.

[38] Galvez, L.; Gonzalez, E. M. and Arrese-Igor, C. (2005). Evidence for carbon flux shortage and strong carbon/nitrogen interactions in pea nodules at early stages of water stress. *J. Exp. Bot.*, *56*, 2551-2561.

[39] Gouia, H.; Suzuki, A.; Brulfert, J. and Ghorbal, M. H. (2003). Effects of cadmium on the co-ordination of nitrogen and carbon metabolism in bean seedlings. *J. Plant Physiol.*, *160*, 367-376.

[40] Popova, O. V.; Popova, T. N. and Izmailov, S. F. (2001). Some features of NADP-dependent isocitrate dehydrogenase functioning in pea leaves upon exposure to salt stress. *Biol. Bull.*, *28*, 134-138.

[41] Molina, C.; Zaman-Allah, M.; Khan, F.; Fatnassi, N.; Horres, R.; Rotter, B.; Steinhauer, D.; Amenc, L.; Drevon, J. J.; Winter, P. and Kahl, G. (2011).

The salt-responsive transcriptome of chickpea roots and nodules *via* deepsupersage. *BMC Plant Biol*, *11*, 1-26.

[42] Saruyama, H. and Tanida, M. (1995). Effect of chilling on activated oxygen-scavenging enzymes in low temperature-sensitive and temperature-tolerant cultivars of rice (*Oryza sativa* L.). *Plant Sci.*, *109*, 105-113.

[43] Lu, B. B.; Yuan, Y. Z.; Zhang, C. F.; Ou, J. Q.; Zhou, W. and Lin, Q. H. (2005). Modulation of key enzymes involved in ammonium assimilation and carbon metabolism by low temperature in rice (*Oryza sativa* L.) roots. *Plant Sci.*, *169*, 295-302.

[44] Lopez-Bucio, J.; de La Vega, O. M.; Guevara-Garcia, A. and Herrera-Estrella, L. (2000). Enhanced phosphorus uptake in transgenic tobacco plants that overproduce citrate. *Nat. Biotechnol.*, *18*, 450-453.

[45] Delhaize, E.; Ryan, P. R.; Hocking, P. J. and Richardson, A. E. (2003). Effects of altered citrate synthase and isocitrate dehydrogenase expression on internal citrate concentrations and citrate efflux from tobacco (*Nicotiana tabacum* L.) roots. *Plant Soil*, *248*, 137-144.

[46] Hayes, J. E. and Ma, J. F. (2003). Al-induced efflux of organic acid anions is poorly associated with internal organic acid metabolism in triticale roots. *J. Exp. Bot.*, *54*, 1753-1759.

[47] Eticha, D.; Zahn, M.; Rao, I. M. and Horst, W. J. (2010). Transcriptomic analysis reveals differential gene expression in response to aluminium in common bean (*Phaseolus vulgaris*) genotypes. *Ann. Bot.*, *105*, 1119-1128.

[48] Guo, P.; Bai, G.; Carver, B.; Li, R.; Bernardo, A. and Baum, M. (2007). Transcriptional analysis between two wheat near-isogenic lines contrasting in aluminum tolerance under aluminum stress. *Mol. Genet. Genomics*, *277*, 1-12.

[49] Drincovich, M. F.; Casati, P. and Andreo, C. S. (2001). NADP-malic enzyme from plants: A ubiquitous enzyme involved in different metabolic pathways. *FEBS Lett.*, *490*, 1-6.

[50] Edwards, G. E. and Andreo, C. S. (1992). NADP-malic enzyme from plants. *Phytochemistry*, *31*, 1845-1857.

[51] Doubnerova, V.; Potuckova, L.; Muller, K. and Ryslava, H. (2009). The regulation and catalytic mechanism of the NADP-malic enzyme from tobacco leaves. *J. Serb. Chem. Soc.*, *74*, 893-906.

[52] Muller, G. L.; Drincovich, M. F.; Andreo, C. S. and Lara, M. V. (2008). *Nicotiana tabacum* NADP-malic enzyme: Cloning, characterization and analysis of biological role. *Plant Cell Physiol.*, *49*, 469-480.

[53] Wheeler, M. C.; Tronconi, M. A.; Drincovich, M. F.; Andreo, C. S.; Flugge, U. I. and Maurino, V. G. (2005). A comprehensive analysis of the NADP-malic enzyme gene family of Arabidopsis. *Plant Physiol.*, *139*, 39-51.

[54] Chi, W.; Yang, J.; Wu, N. and Zhang, F. (2004). Four rice genes encoding NADP malic enzyme exhibit distinct expression profiles. *Biosci. Biotechnol. Biochem.*, *68*, 1865-1874.

[55] Doubnerova, V. and Ryslava, H. (2011). What can enzymes of $C_4$ photosynthesis do for $C_3$ plants under stress? *Plant Sci.*, *180*, 575-583.

[56] Doubnerova, V.; Muller, K.; Cerovska, N.; Synkova, H.; Spoustova, P. and Ryslava, H. (2009). Effect of *Potato virus Y* on the NADP-malic enzyme from *Nicotiana tabacum* L.: mRNA, expressed protein and activity. *Int. J. Mol. Sci.*, *10*, 3583-3598.

[57] Ryslava, H.; Muller, K.; Semoradova, S.; Synkova, H. and Cerovska, N. (2003). Photosynthesis and activity of phosphoenolpyruvate carboxylase in *Nicotiana tabacum* L. leaves infected by *Potato virus A* and *Potato virus Y*. *Photosynthetica*, *41*, 357-363.

[58] Liu, S.; Cheng, Y.; Zhang, X.; Guan, Q.; Nishiuchi, S.; Hase, K. and Takano, T. (2007). Expression of an NADP-malic enzyme gene in rice (*Oryza sativa* L.) is induced by environmental stresses; over-expression of the gene in Arabidopsis confers salt and osmotic stress tolerance. *Plant Mol. Biol.*, *64*, 49-58.

[59] Guo, P.; Baum, M.; Grando, S.; Ceccarelli, S.; Bai, G.; Li, R.; von Korff, M.; Varshney, R. K.; Graner, A. and Valkoun, J. (2009). Differentially expressed genes between drought-tolerant and drought-sensitive barley genotypes in response to drought stress during the reproductive stage. *J. Exp. Bot.*, *60*, 3531-3544.

[60] Dizengremel, P.; Le Thiec, D.; Bagard, M. and Jolivet, Y. (2008). Ozone risk assessment for plants: Central role of metabolism-dependent changes in reducing power. *Environ. Pollut.*, *156*, 11-15.

[61] Dizengremel, P.; Le Thiec, D.; Hasenfratz-Sauder, M. P.; Vaultier, M. N.; Bagard, M. and Jolivet, Y. (2009). Metabolic-dependent changes in plant cell redox power after ozone exposure. *Plant Biol. (Stuttg)*, *11 Suppl 1*, 35-42.

[62] Fu, Z. Y.; Zhang, Z. B.; Liu, Z. H.; Hu, X. J. and Xu, P. (2011). The effects of abiotic stresses on the NADP-dependent malic enzyme in the leaves of the hexaploid wheat. *Biol. Plant.*, *55*, 196-200.

[63] Laporte, M. M.; Shen, B. and Tarczynski, M. C. (2002). Engineering for drought avoidance: Expression of maize NADP-malic enzyme in tobacco results in altered stomatal function. *J. Exp. Bot.*, *53*, 699-705.

[64] Ameziane, R.; Bernhard, K. and Lightfoot, D. (2000). Expression of the bacterial *gdhA* gene encoding a NADPH glutamate dehydrogenase in tobacco affects plant growth and development. *Plant Soil*, *221*, 47-57.

[65] de-Bashan, L. E.; Magallon, P.; Antoun, H. and Bashan, Y. (2008). Role of glutamate dehydrogenase and glutamine synthetase in *Chlorella vulgaris* during assimilation of ammonium when jointly immobilized with the microalgae-growth-promoting bacterium *Azospirillum brasilense*. *J. Phycol. 44*, 1188-1196.

[66] Kisaka, H. and Kida, T. (2003). Transgenic tomato plant carrying a gene for NADP-dependent glutamate dehydrogenase (*gdhA*) from *Aspergillus nidulans*. *Plant Sci.*, *164*, 35-42.

[67] Lightfoot, D. A.; Mungur, R.; Ameziane, R.; Nolte, S.; Long, L.; Bernhard, K.; Colter, A.; Jones, K.; Iqbal, M. J.; Varsa, E. and Young, B. (2007). Improved drought tolerance of transgenic *Zea mays* plants that express the glutamate dehydrogenase gene (*gdhA*) of *E. coli*. *Euphytica*, *156*, 103-116.

[68] Kakinuma, M.; Coury, D. A.; Kuno, Y.; Itoh, S.; Kozawa, Y.; Inagaki, E.; Yoshiura, Y. and Amano, H. (2006). Physiological and biochemical responses to thermal and salinity stresses in a sterile mutant of *Ulva pertusa* (Ulvales, chlorophyta). *Mar. Biol.*, *149*, 97-106.

[69] Kakinuma, M.; Shibahara, N.; Ikeda, H.; Maegawa, M. and Amano, H. (2001). Thermal stress responses of a sterile mutant of *Ulva pertusa* (chlorophyta). *Fisher. Sci.*, *67*, 287-294.

[70] Mungur, R.; Wood, A. J. and Lightfoot, D. A. (2006). Water potential is maintained during water deficit in *Nicotiana tabacum* expressing the *Escherichia coli* glutamate dehydrogenase gene. *Plant Growth Reg.*, *50*, 231-238.

[71] Bustos, D. M. and Iglesias, A. A. (2002). Non-phosphorylating glyceraldehyde-3-phosphate dehydrogenase is post-translationally phosphorylated in heterotrophic cells of wheat (*Triticum aestivum*). *FEBS Lett.*, *530*, 169-173.

[72] Kirch, H. H.; Bartels, D.; Wei, Y. L.; Schnable, P. S. and Wood, A. J. (2004). The ALDH gene superfamily of Arabidopsis. *Trends Plant Sci.*, *9*, 371-377.

[73] Bustos, D. M.; Bustamante, C. A. and Iglesias, A. A. (2008). Involvement of non-phosphorylating glyceraldehyde-3-phosphate dehydrogenase in response to oxidative stress. *J. Plant Physiol.*, *165*, 456-461.

[74] Gao, Z. and Loescher, W. H. (2000). NADPH supply and mannitol biosynthesis. Characterization, cloning, and regulation of the non-reversible

glyceraldehyde-3-phosphate dehydrogenase in celery leaves. *Plant Physiol.*, *124*, 321-330.

[75] Rius, S. P.; Casati, P.; Iglesias, A. A. and Gomez-Casati, D. F. (2006). Characterization of an *Arabidopsis thaliana* mutant lacking a cytosolic non-phosphorylating glyceraldehyde-3-phosphate dehydrogenase. *Plant Mol. Biol.*, *61*, 945-957.

[76] Ferreyra, M. L. F.; Perotti, V.; Figueroa, C. M.; Garran, S.; Anderson, P. C.; Vazquez, D.; Iglesias, A. A. and Podesta, F. E. (2006). Carbohydrate metabolism and fruit quality are affected in frost-exposed Valencia orange fruit. *Physiol. Plant.*, *128*, 224-236.

[77] Pillai, M. A.; Lihuang, Z. and Akiyama, T. (2002). Molecular cloning, characterization, expression and chromosomal location of osgapdh, a submergence responsive gene in rice (*Oryza sativa* L.). *Theor. App. Genet.*, *105*, 34-42.

[78] Shenoy, V. V. and Kalagudi, G. M. (2005). Enhancing plant phosphorus use efficiency for sustainable cropping. *Biotechnol. Adv.*, *23*, 501-513.

[79] Duff, S. M.; Moorhead, G. B.; Lefebvre, D. D. and Plaxton, W. C. (1989). Phosphate starvation inducible "bypasses" of adenylate and phosphate dependent glycolytic enzymes in *Brassica nigra* suspension cells. *Plant Physiol.*, *90*, 1275-1278.

[80] Shinano, T.; Nanamori, M.; Dohi, M.; Wasaki, J. and Osaki, M. (2005). Evaluation of phosphorus starvation inducible genes relating to efficient phosphorus utilization in rice. *Plant Soil*, *269*, 81-87.

[81] Ding, L.; Hofius, D.; Hajirezaei, M. R.; Fernie, A. R.; Bornke, F. and Sonnewald, U. (2007). Functional analysis of the essential bifunctional tobacco enzyme 3-dehydroquinate dehydratase/shikimate dehydrogenase in transgenic tobacco plants. *J. Exp. Bot.*, *58*, 2053-2067.

[82] Tzin, V. and Galili, G. (2010). New insights into the shikimate and aromatic amino acids biosynthesis pathways in plants. *Mol Plant*, *3*, 956-972.

[83] Pagnussat, G. C.; Yu, H. J.; Ngo, Q. A.; Rajani, S.; Mayalagu, S.; Johnson, C. S.; Capron, A.; Xie, L. F.; Ye, D. and Sundaresan, V. (2005). Genetic and molecular identification of genes required for female gametophyte development and function in Arabidopsis *Development*, *132*, 1161-1161.

[84] Heldt, H.-W.; Piechulla, B. and Heldt, F. (2011). *Plant Biochemistry-translation of the 4$^{th}$ German edition* (4$^{th}$). Amsterdam, Boston, Heildelberg, London, New York, Oxford, Paris, San Diego, San Francisco, Singapore, Sydney, Tokyo: Elsevier Academic Press.

[85] Betz, G. A.; Knappe, C.; Lapierre, C.; Olbrich, M.; Welzl, G.; Langebartels, C.; Heller, W.; Sandermann, H. and Ernst, D. (2009). Ozone affects

shikimate pathway transcripts and monomeric lignin composition in European beech (*Fagus sylvatica* L.). *European J. Forest Res.*, *128*, 109-116.

[86] Marchiosi, R.; Ferrarese, M. D. L. L.; Bonini, E. A.; Fernandes, N. G.; Ferro, A. P. and Ferrarese-Filho, O. (2009). Glyphosate-induced metabolic changes in susceptible and glyphosate-resistant soybean (*Glycine max* l.) roots. *Pestic. Biochem. Phys. 93*, 28-33.

[87] Wang, C.; Lu, J.; Zhang, S.; Wang, P.; Hou, J. and Qian, J. (2011). Effects of Pb stress on nutrient uptake and secondary metabolism in submerged macrophyte *Vallisneria natans*. *Ecotoxicol. Environ. Saf.*, *74*, 1297-1303.

[88] Kovacik, J.; Klejdus, B. and Backor, M. (2009). Phenolic metabolism of *Matricaria chamomilla* plants exposed to nickel. *J. Plant Physiol.*, *166*, 1460-1464.

[89] Kovacik, J.; Klejdus, B.; Hedbavny, J. and Backor, M. (2010). Tolerance of *Silene vulgaris* to copper: Population-related comparison of selected physiological parameters. *Environ. Toxicol.*, *25*, 581-592.

[90] Diaz, J.; Bernal, A.; Pomar, F. and Merino, F. (2001). Induction of shikimate dehydrogenase and peroxidase in pepper (*Capsicum annuum* L.) seedlings in response to copper stress and its relation to lignification. *Plant Sci.*, *161*, 179-188.

[91] Moura, J. C.; Bonine, C. A.; de Oliveira Fernandes Viana, J.; Dornelas, M. C. and Mazzafera, P. (2010). Abiotic and biotic stresses and changes in the lignin content and composition in plants. *J. Integr. Plant Biol.*, *52*, 360-376.

[92] Ali, M. B.; Hahn, E. J. and Paek, K. Y. (2007). Methyl jasmonate and salicylic acid induced oxidative stress and accumulation of phenolics in *Panax ginseng* bioreactor root suspension cultures. *Molecules*, *12*, 607-621.

[93] Cao, X.; Zong, Z.; Ju, X.; Sun, Y.; Dai, C.; Liu, Q. and Jiang, J. (2010). Molecular cloning, characterization and function analysis of the gene encoding HMG-CoA reductase from *Euphorbia pekinensis* Rupr. *Mol. Biol. Rep.*, *37*, 1559-1567.

[94] Everard, J. D.; Gucci, R.; Kann, S. C.; Flore, J. A. and Loescher, W. H. (1994). Gas-exchange and carbon partitioning in the leaves of celery (*Apium graveolens* L.) at various levels of root-zone salinity. *Plant Physiol.*, *106*, 281-292.

[95] Everard, J. D.; Cantini, C.; Grumet, R.; Plummer, J. and Loescher, W. H. (1997). Molecular cloning of mannose-6-phosphate reductase and its developmental expression in celery. *Plant Physiol.*, *113*, 1427-1435.

[96] Dichio, B.; Margiotta, G.; Xiloyannis, C.; Bufo, S. A.; Sofo, A. and Cataldi, T. R. I. (2009). Changes in water status and osmolyte contents in leaves and

roots of olive plants (*Olea europaea* L.) subjected to water deficit. *Trees-Struct. Funct. 23*, 247-256.

[97] Partelli, F. L.; Vieira, H. D.; Rodrigues, A. P. D.; Pais, I.; Campostrini, E.; Chaves, M. M. C. C. and Ramalho, J. C. (2010). Cold induced changes on sugar contents and respiratory enzyme activities in coffee genotypes. *Cienc. Rural, 40*, 781-786.

[98] Stoop, J. and Pharr, D. M. (1994). Mannitol metabolism in celery stressed by excess macronutrients. *Plant Physiol., 106*, 503-511.

[99] Chan, Z.; Grumet, R. and Loescher, W. (2011). Global gene expression analysis of transgenic, mannitol-producing, and salt-tolerant *Arabidopsis thaliana* indicates widespread changes in abiotic and biotic stress-related genes. *J. Exp. Bot., 62*, 4782-4803.

[100] Manimaran, P.; Ramkumar, G.; Sakthivel, K.; Sundaram, R. M.; Madhav, M. S. and Balachandran, S. M. (2011). Suitability of non-lethal marker and marker-free systems for development of transgenic crop plants: Present status and future prospects. *Biotechnol. Adv., 29*, 703-704.

[101] Aly, R.; Cholakh, H.; Joel, D. M.; Leibman, D.; Steinitz, B.; Zelcer, A.; Naglis, A.; Yarden, O. and Gal-On, A. (2009). Gene silencing of mannose 6-phosphate reductase in the parasitic weed *Orobanche aegyptiaca* through the production of homologous dsRNA sequences in the host plant. *Plant Biotechnol. J., 7*, 487-498.

[102] Kishor, P. B. K.; Sangam, S.; Amrutha, R. N.; Sri Laxmi, P.; Naidu, K. R.; Rao, K. R. S. S.; Rao, S.; Reddy, K. J.; Theriappan, P. and Sreenivasulu, N. (2005). Regulation of proline biosynthesis, degradation, uptake and transport in higher plants: its implications in plant growth and abiotic stress tolerance. *Curr. Sci., 88*, 424-438.

[103] Lehmann, S.; Funck, D.; Szabados, L. and Rentsch, D. (2010). Proline metabolism and transport in plant development. *Amino Acids, 39*, 949-962.

[104] Perez-Arellano, I.; Carmona-Alvarez, F.; Martinez, A. I.; Rodriguez-Diaz, J. and Cervera, J. (2010). Pyrroline-5-carboxylate synthase and proline biosynthesis: from osmotolerance to rare metabolic disease. *Protein Sci., 19*, 372-382.

[105] Liu, C.; Zhao, L. and Yu, G. (2011). The dominant glutamic acid metabolic flux to produce gamma-amino butyric acid over proline in *Nicotiana tabacum* leaves under water stress relates to its significant role in antioxidant activity. *J. Integr. Plant Biol., 53*, 608-618.

[106] Parre, E.; de Virville, J.; Cochet, F.; Leprince, A. S.; Richard, L.; Lefebvre-De Vos, D.; Ghars, M. A.; Bordenave, M.; Zachowski, A. and Savoure, A.

(2010). A new method for accurately measuring delta[1]-pyrroline-5-carboxylate synthetase activity. *Methods Mol. Biol.*, *639*, 333-340.

[107] Szekely, G.; Abraham, E.; Cseplo, A.; Rigo, G.; Zsigmond, L.; Csiszar, J.; Ayaydin, F.; Strizhov, N.; Jasik, J.; Schmelzer, E.; Koncz, C. and Szabados, L. (2008). Duplicated P5CS genes of Arabidopsis play distinct roles in stress regulation and developmental control of proline biosynthesis. *Plant J.*, *53*, 11-28.

[108] Mattioli, R.; Costantino, P. and Trovato, M. (2009). Proline accumulation in plants: not only stress. *Plant Signal. Behav.*, *4*, 1016-1018.

[109] Kishor, P. B. K.; Hong, Z. L.; Miao, G. H.; Hu, Ch.-A. A. and Verma, D. P. S. (1995). Overexpression of delta-pyrroline-5-carboxylate synthetase increases proline production and confers osmotolerance in transgenic plants. *Plant Physiol.*, *108*, 1387-1394.

[110] Wang, H.; Nagegowda, D. A.; Rawat, R.; Bouvier-Nave, P.; Guo, D.; Bach, T. J. and Chye, M. L. (2011). Overexpression of *Brassica juncea* wild-type and mutant HMG-CoA synthase 1 in Arabidopsis up-regulates genes in sterol biosynthesis and enhances sterol production and stress tolerance. *Plant Biotechnol. J.*, *doi: 10.1111/j.1467-7652.2011.00631*, 1.

[111] Kondo, K.; Uritani, I. and Oba, K. (2003). Induction mechanism of 3-hydroxy-3-methylglutaryl-CoA reductase in potato tuber and sweet potato root tissues. *Biosci. Biotechnol. Biochem.*, *67*, 1007-1017.

[112] Suzuki, M.; Kamide, Y.; Nagata, N.; Seki, H.; Ohyama, K.; Kato, H.; Masuda, K.; Sato, S.; Kato, T.; Tabata, S.; Yoshida, S. and Muranaka, T. (2004). Loss of function of 3-hydroxy-3-methylglutaryl coenzyme A reductase 1 (HMG1) in Arabidopsis leads to dwarfing, early senescence and male sterility, and reduced sterol levels. *Plant J.*, *37*, 750-761.

[113] Munoz-Bertomeu, J.; Sales, E.; Ros, R.; Arrillaga, I. and Segura, J. (2007). Up-regulation of an N-terminal truncated 3-hydroxy-3-methylglutaryl CoA reductase enhances production of essential oils and sterols in transgenic *Lavandula latifolia*. *Plant Biotechnol. J.*, *5*, 746-758.

[114] Leivar, P.; Gonzalez, V. M.; Castel, S.; Trelease, R. N.; Lopez-Iglesias, C.; Arro, M.; Boronat, A.; Campos, N.; Ferrer, A. and Fernandez-Busquets, X. (2005). Subcellular localization of Arabidopsis 3-hydroxy-3-methylglutaryl-coenzyme A reductase. *Plant Physiol.*, *137*, 57-69.

[115] Holmberg, N.; Harker, M.; Wallace, A. D.; Clayton, J. C.; Gibbard, C. L. and Safford, R. (2003). Co-expression of N-terminal truncated 3-hydroxy-3-methylglutaryl CoA reductase and C24-sterol methyltransferase type 1 in transgenic tobacco enhances carbon flux towards end-product sterols. *Plant J.*, *36*, 12-20.

[116] Chen, C. J.; Huang, C. Y.; Huang, J. K.; Lin, C. Y. and Lin, C. T. (2009). Cloning, expression, and purification of a functional glutathione reductase from sweet potato (*Ipomoea batatas* L. Lam): Kinetic studies and characterization. *J. Agric. Food Chem.*, *57*, 4403-4408.

[117] Contour-Ansel, D.; Torres-Franklin, M. L.; Cruz De Carvalho, M. H.; D'Arcy-Lameta, A. and Zuily-Fodil, Y. (2006). Glutathione reductase in leaves of cowpea: cloning of two cDNAs, expression and enzymatic activity under progressive drought stress, desiccation and abscisic acid treatment. *Ann. Bot.*, *98*, 1279-1287.

[118] Hong, C. Y.; Chao, Y. Y.; Yang, M. Y.; Cheng, S. Y.; Cho, S. C. and Kao, C. H. (2009). NaCl-induced expression of glutathione reductase in roots of rice (*Oryza sativa* L.) seedlings is mediated through hydrogen peroxide but not abscisic acid. *Plant Soil*, *320*, 103-115.

[119] Hong, C. Y.; Chao, Y. Y.; Yang, M. Y.; Cho, S. C. and Huei Kao, C. (2009). $Na^+$ but not $Cl^-$ or osmotic stress is involved in NaCl-induced expression of glutathione reductase in roots of rice seedlings. *J. Plant Physiol.*, *166*, 1598-1606.

[120] Murthy, S. S. and Zilinskas, B. A. (1994). Molecular cloning and characterization of a cDNA encoding pea monodehydroascorbate reductase. *J. Biol. Chem.*, *269*, 31129-31133.

[121] Yoon, H. S.; Lee, H.; Lee, I. A.; Kim, K. Y. and Jo, J. (2004). Molecular cloning of the monodehydroascorbate reductase gene from *Brassica campestris* and analysis of its mRNA level in response to oxidative stress. *Biochim. Biophys. Acta*, *1658*, 181-186.

[122] Igamberdiev, A. U.; Bykova, N. V. and Hill, R. D. (2006). Nitric oxide scavenging by barley hemoglobin is facilitated by a monodehydroascorbate reductase-mediated ascorbate reduction of methemoglobin. *Planta*, *223*, 1033-1040.

[123] Leterrier, M.; Corpas, F. J.; Barroso, J. B.; Sandalio, L. M. and del Rio, L. A. (2005). Peroxisomal monodehydroascorbate reductase. Genomic clone characterization and functional analysis under environmental stress conditions. *Plant Physiol.*, *138*, 2111-2123.

[124] Eltelib, H. A.; Badejo, A. A.; Fujikawa, Y. and Esaka, M. (2011). Gene expression of monodehydroascorbate reductase and dehydroascorbate reductase during fruit ripening and in response to environmental stresses in acerola (*Malpighia glabra*). *J. Plant Physiol.*, *168*, 619-627.

[125] Eltayeb, A. E.; Kawano, N.; Badawi, G. H.; Kaminaka, H.; Sanekata, T.; Shibahara, T.; Inanaga, S. and Tanaka, K. (2007). Overexpression of monodehydroascorbate reductase in transgenic tobacco confers enhanced

tolerance to ozone, salt and polyethylene glycol stresses. *Planta*, *225*, 1255-1264.

[126] Stevens, R.; Page, D.; Gouble, B.; Garchery, C.; Zamir, D. and Causse, M. (2008). Tomato fruit ascorbic acid content is linked with monodehydroascorbate reductase activity and tolerance to chilling stress. *Plant Cell Environ.*, *31*, 1086-1096.

[127] Li, F.; Wu, Q. Y.; Sun, Y. L.; Wang, L. Y.; Yang, X. H. and Meng, Q. W. (2010). Overexpression of chloroplastic monodehydroascorbate reductase enhanced tolerance to temperature and methyl viologen-mediated oxidative stresses. *Physiol. Plant.*, *139*, 421-434.

[128] Secenji, M.; Hideg, E.; Bebes, A. and Gyorgyey, J. (2010). Transcriptional differences in gene families of the ascorbate-glutathione cycle in wheat during mild water deficit. *Plant Cell Rep.*, *29*, 37-50.

[129] Markovska, Y. K.; Gorinova, N. I.; Nedkovska, M. P. and Miteva, K. M. (2009). Cadmium-induced oxidative damage and antioxidant responses in *Brassica juncea* plants. *Biol. Plant.*, *53*, 151-154.

[130] Sanchez-Rodriguez, E.; Rubio-Wilhelmi, M.; Cervilla, L. M.; Blasco, B.; Rios, J. J.; Rosales, M. A.; Romero, L. and Ruiz, J. M. (2010). Genotypic differences in some physiological parameters symptomatic for oxidative stress under moderate drought in tomato plants. *Plant Sci.*, *178*, 30-40.

[131] Lu, P.; Sang, W. G. and Ma, K. P. (2008). Differential responses of the activities of antioxidant enzymes to thermal stresses between two invasive *Eupatorium* species in China. *J. Integr. Plant Biol.*, *50*, 393-401.

[132] Bai, R.; Ma, F.; Liang, D. and Zhao, X. (2009). Phthalic acid induces oxidative stress and alters the activity of some antioxidant enzymes in roots of *Malus prunifolia*. *J. Chem. Ecol.*, *35*, 488-494.

[133] Yannarelli, G. G.; Fernandez-Alvarez, A. J.; Santa-Cruz, D. M. and Tomaro, M. L. (2007). Glutathione reductase activity and isoforms in leaves and roots of wheat plants subjected to cadmium stress. *Phytochemistry*, *68*, 505-512.

[134] Ding, S.; Lu, Q.; Zhang, Y.; Yang, Z.; Wen, X.; Zhang, L. and Lu, C. (2009). Enhanced sensitivity to oxidative stress in transgenic tobacco plants with decreased glutathione reductase activity leads to a decrease in ascorbate pool and ascorbate redox state. *Plant Mol. Biol.*, *69*, 577-592.

[135] de Lamotte, F.; Vianey-Liaud, N.; Duviau, M. P. and Kobrehel, K. (2000). Glutathione reductase in wheat grain. 1. Isolation and characterization. *J. Agricul. Food Chem.*, *48*, 4978-4983.

[136] Meloni, D. A.; Oliva, M. A.; Martinez, C. A. and Cambraia, J. (2003). Photosynthesis and activity of superoxide dismutase, peroxidase and

glutathione reductase in cotton under salt stress. *Environ. Exp. Bot.*, *49*, 69-76.

[137] Rao, M. V.; Paliyath, C. and Ormrod, D. P. (1996). Ultraviolet-B- and ozone-induced biochemical changes in antioxidant enzymes of *Arabidopsis thaliana. Plant Physiol.*, *110*, 125-136.

[138] Kidambi, S. P.; Mahan, J. R. and Matches, A. G. (1990). Purification and thermal dependence of glutathione reductase from two forage legume species. *Plant Physiol.*, *92*, 363-367.

[139] Edwards, E. A.; Enard, C.; Creissen, G. P. and Mullineaux, P. M. (1994). Synthesis and properties of glutathione reductase in stressed peas. *Planta*, *192*, 137-143.

[140] Serrato, A. J.; Perez-Ruiz, J. M.; Spinola, M. C. and Cejudo, F. J. (2004). A novel NADPH thioredoxin reductase, localized in the chloroplast, which deficiency causes hypersensitivity to abiotic stress in *Arabidopsis thaliana. J. Biol. Chem.*, *279*, 43821-43827.

[141] Montrichard, F.; Alkhalfioui, F.; Yano, H.; Vensel, W. H.; Hurkman, W. J. and Buchanan, B. B. (2009). Thioredoxin targets in plants: The first 30 years. *J. Proteomics*, *72*, 452-474.

[142] Buchanan, B. B. and Balmer, Y. (2003). Plant thioredoxins: Role of the French school. *Plant Physiol. Biochem.*, *41*, 505-512.

[143] Perez-Ruiz, J. M.; Spinola, M. C.; Kirchsteiger, K.; Moreno, J.; Sahrawy, M. and Cejudo, F. J. (2006). Rice NTRC is a high-efficiency redox system for chloroplast protection against oxidative damage. *Plant Cell*, *18*, 2356-2368.

[144] Spinola, M. C.; Perez-Ruiz, J. M.; Pulido, P.; Kirchsteiger, K.; Guinea, M.; Gonzalez, M. and Cejudo, F. J. (2008). NTRC new ways of using NADPH in the chloroplast. *Physiol. Plant.*, *133*, 516-524.

[145] Meyer, Y.; Siala, W.; Bashandy, T.; Riondet, C.; Vignols, F. and Reichheld, J. P. (2008). Glutaredoxins and thioredoxins in plants. *Biochim. Biophys. Acta*, *1783*, 589-600.

[146] Meyer, Y.; Reichheld, J. P. and Vignols, F. (2005). Thioredoxins in Arabidopsis and other plants. *Photosynth. Res.*, *86*, 419-433.

[147] Buchanan, B. B. and Luan, S. (2005). Redox regulation in the chloroplast thylakoid lumen: a new frontier in photosynthesis research. *J. Exp. Bot.*, *56*, 1439-1447.

In: Abiotic Stress: New Research
Editors: Nikhil Haryana and Shreya Punj
ISBN: 978-1- 61942-197-4


*Chapter 3*

# WATER-DEFICIT STRESS SIGNAL TRANSDUCTION PATHWAYS IN PLANTS: FROM SENSING TO RESPONSE

Daniela A. Capiati*, María Noelia Muñiz García and ***Rita M. Ulloa***

Instituto de Investigaciones en Ingeniería Genética y Biología Molecular, INGEBI, CONICET and Facultad de Ciencias Exactas y Naturales, Universidad de Buenos Aires, Buenos Aires, Argentina

## ABSTRACT

As sessile organisms plants have to cope with changing environmental conditions. Drought and salinity, which causes water-deficit in plant cells, are common adverse factors that limit plant growth and productivity. Understanding the mechanisms by which plants perceive environmental signals and transmit them to cellular machinery to activate adaptive responses is of great importance to biology and to rational engineering of crop plants. This chapter reviews the signal transduction mechanisms that activate water-deficit stress responses and the regulation of transcription factors that control the expression of stress-responsive genes. The general components of stress signal transduction pathway for drought and salt stress are considered. Signal perception, receptor-coupled phosphorelay,

* Corresponding author: TE: 0054-11-4783-2871, FAX: 0054-11-4786-8576, Email: dcapiati@dna.uba.ar; dcapiati@gmail.com

phosphoinositol-induced $Ca^{2+}$ changes, $Ca^{2+}$-coupled phosphoprotein cascades, mitogen-activated protein kinase cascade and transcriptional activation of stress responsive genes are the main signal transduction steps addressed. Abscisic acid (ABA) plays a pivotal role in stress responses in plants. Therefore, the hormone implications are also reviewed. The transcription factors responsible for reprogramming gene expression in response to stress are described. Finally, the physiological and biochemical responses that lead to plant tolerance to water-deficit stress are addressed.

## INTRODUCTION

Abiotic stress responses are important for sessile organisms such as plants because this type of organisms cannot survive unless they are able to cope with environmental changes. Water- deficit stresses are common abiotic conditions that adversely affect plant growth and crop production. Therefore, understanding the mechanisms by which plants perceive and transduce these stress signals to activate adaptive responses is essential for the development of transgenic strategies to improve water-deficit stress tolerance in crops.

Many environmental conditions can lead to water deficit. Drought caused by little or no rainfall reduces the amount of water available for plant growth. Nevertheless, water deficit may occur in environments in which water is not limiting, for instance, saline habitats. The presence of high salt concentrations in the soil makes it more difficult for roots to extract water. In addition to water deficit due to osmotic stress, plants growing under saline conditions are exposed to toxic concentrations of specific ions.

The cellular and molecular responses of plants to drought and salt stress have been studied intensively [1-4]. In response to water-deficit stress, various genes are up-regulated and their products can mitigate the effect of stress leading to the adjustment of the cellular milieu and plant tolerance. In this chapter, we consider common characteristics of water-deficit stress signal transduction pathways and the regulation of transcription factors that control the expression of stress-responsive genes. The physiological and biochemical responses that improve tolerance to water-deficit stress are also addressed.

# WATER-DEFICIT STRESS SIGNAL TRANSDUCTION PATHWAYS

Signal transduction pathways start with signal perception, followed by the generation of second messengers such as $Ca^{2+}$, inositol phosphates and reactive oxygen species (ROS). Second messengers can activate protein phosphorylation cascades regulated by kinases like CDPKs ($Ca^{2+}$-dependent protein kinases), PKSs (SOS2-like protein kinases) and MAPKs (mitogen activated protein kinases), and protein phosphatases. These phosphorylation cascades finally target transcription factors that control sets of stress-responsive genes, whose products may contribute to mitigate the cellular damage caused by stress, like antioxidants or osmolytes synthesizing enzymes and late embryogenesis abundant (LEA) proteins (Figure 1). The products of stress-responsive genes may also participate

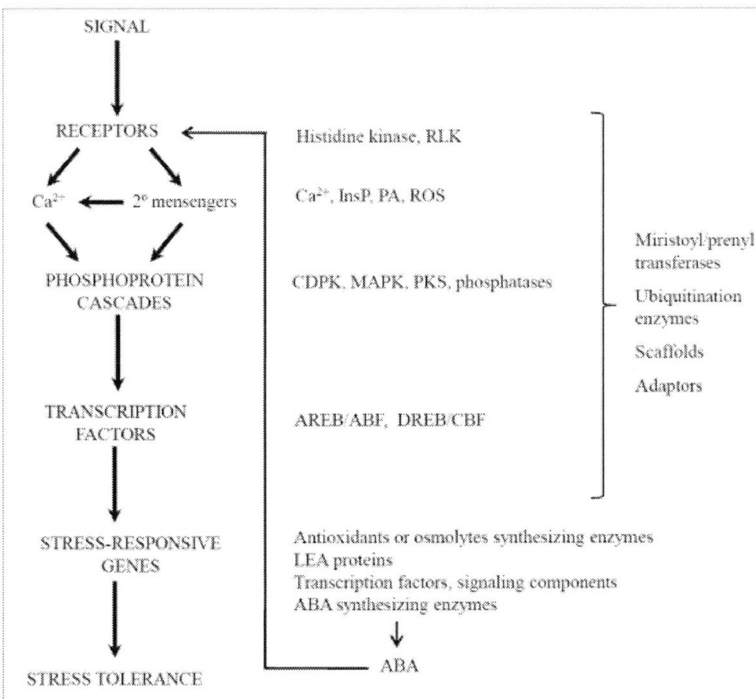


Figure 1. Signal transduction pathways of water-stress deficit (adapted from Xiong et al. [5]). RLK, receptor-like protein kinase; InsP, inositol polyphosphates; PA, phosphatidic acid; ABF/AREB, ABRE-binding factor/ABA-response element binding factor. Other abbreviations are given in the text.

in the generation of regulatory molecules like the hormone abscisic acid (ABA), which plays a crucial role to integrate plant response to water-deficit stress. ABA can initiate a second round of signaling that may follow the above generic pathway, although different components are involved (Figure 1).

Certain proteins involved in the modification, delivery, or assembly of signaling components are also important in stress response signal transduction. These molecules include protein modifiers (enzymes for protein lipidation, methylation, glycosylation, and ubiquitination), scaffolds and adaptors [5]. They are critical for the accurate transmission of stress signals since they are required for the proper spatial and temporal coordination of all signaling components.

## POTENTIAL SENSORS FOR WATER-DEFICIT STRESS

As all living organisms, plant cells perceive and process information through receptors exposed on the cell surface. Membrane protein kinases of two classes may act as receptors in osmotic stress: receptor His kinases (HKs) and receptor-like protein kinases (RLKs).

Two-component systems, consisting of a sensory histidine kinase and a response regulator, function as stress sensors in bacteria and yeast. For example, the transmembrane two-component histidine kinases HIK33 (cyanobacterium) and DesK (*Bacillus subtilis*) have been suggested as thermosensors [6, 7]. In yeast cells, the osmosensory histidine kinase SLN1 senses the changes in osmolarity and triggers a two-component system, which activates the HOG1 (high-osmolarity glycerol response 1) MAPK cascade [8]. The *Arabidopsis thaliana* histidine kinase ATHK1, is thought to be a candidate osmosensor, since it is induced by salt stress (250 mM NaCl) and is able to complement the yeast double mutant *sln1*Δ *sho1*Δ, defective in osmosensing, demonstrating that ATHK1 can activate the HOG1 MAPK pathway in yeast [9].

NtC7 (*Nicotiana tabacum*), a receptor-like protein has been implicated in osmotic stress sensing [10]. *NtC7* is induced by salt (200 mM NaCl) and osmotic stress (500 mM mannitol) in tobacco. Transgenic overexpression of *NtC7* in tobacco plants increments osmotic stress tolerance. Seeds of transgenic tobacco overexpressing *NtC7* normally germinated and grew in the presence of 500 mm mannitol, but not in the presence of 220 mm NaCl or 60 mm LiCl. These results suggest the NtC7 function in osmotic stress is not related to the maintenance of ion homeostasis. *NtC7* codes for a transmembrane protein of 308 amino acid residues. Its N-terminal region is similar to the RLK receptor domain, but NtC7 lacks the kinase domain exposed into the cytoplasm, therefore, the mechanism by which NtC7 transmits the signal to the downstream signaling components remains unknown. NtC7 is homologous to the tomato Cf-9, a transmembrane receptor of the *Cladosporium fulvum* avr-9 elicitor. In the cytoplasm, Cf-9 probably interacts with Ser/Thr kinase Pto, which is involved in signal transduction upon a pathogen attack. It is logical to speculate that NtC7 similarly interacts with its partners in the cytoplasm to transmit the signal in response to osmotic stress.

Recently, a phosphoinositide-specific phospholipase C, designated NtPI-PLC, was identified as an interaction partner of NtC7 by the yeast two-hybrid screening [11]. PI-PLC is a key enzyme in metabolism of inositol phospholipids which function in osmotic stress response [12]. It has been proposed that one of the biological functions of NtC7 is to serve as an anchor for Pi-PLC in the plasma membrane. Therefore, overexpression of NtC7 in tobacco plants confers osmotic stress tolerance probably by tethering more PI-PLC to the plasma membrane and thus strengthening its function.

## Second Messengers

### Calcium

An early response to drought and salt stress is a transient increase in cytosolic $Ca^{2+}$ derived from influx from the apoplastic space or release from internal stores (endoplasmic reticulum, *vacuole*, mitochondria [13, 14]). Internal $Ca^{2+}$ release is regulated by ligand-activated $Ca^{2+}$ channels. These ligands are second messengers such as inositol polyphosphates, cyclic ADP ribose, and nicotinic acid adenine dinucleotide phosphate [15]. A distinctive feature of $Ca^{2+}$ as a signal is the presence of repetitive transient increases in cytosolic $Ca^{2+}$ concentration. These $Ca^{2+}$ transients are triggered by second messengers activated by stress and by ABA generated as a result of cascades activated by early $Ca^{2+}$ signals. These rounds of $Ca^{2+}$ oscillations lead to activation of different signaling cascades with different physiological results.

### Phospholipids

During exposure to stress, phospholipids of the plasma membranes function as precursors for the generation of second-messenger molecules. The phosphoinositide-specific phospholipase C (PI-PLC) hydrolyzes phosphatidylinositol 4,5-bisphosphate ($PIP_2$) upon activation. In response to osmotic stress, plant cells increase the generation of $PIP_2$ [16, 17] by inducing the expression of the *PI5K* gene, which encodes a phosphatidylinositol 4-phosphate 5-kinase [18] responsible for the production of $PIP_2$. Drought or salt stress increases the expression of certain PI-PLC isoforms [19, 20], leading to enhanced cleavage of $PIP_2$ that produces diacylglycerol and inositol 1,4,5-trisphosphate ($IP_3$). $IP_3$ is a second messenger that triggers $Ca^{2+}$ release from intracellular stores [14, 15].

The involvement of $IP_3$ in salt stress signaling was first addressed by the genetic analysis of the FRY1 locus of Arabidopsis. FRY1 encodes an inositol polyphosphate 1-phosphatase, which catabolizes $IP_3$. The *fry1* mutant shows a more sustained accumulation of $IP_3$ in response to ABA, however, it is hypersensitive to salt stress [21], suggesting that $IP_3$ may play a negative role in the salt stress adaptive response. Supporting this hypothesis, transgenic Arabidopsis plants expressing the mammalian type I inositol polyphosphate 5-phosphatase (InsP 5-ptase), which specifically hydrolyzes inositol phosphates, thus terminating the signal, exhibit increased drought tolerance [22]. Drought-inducible ABA-independent transcription factor DREB2A and a subset of DREB2A-regulated genes are up-regulated in the InsP 5-ptase plants, suggesting that $IP_3$ is a negative regulator of the DREB2A drought signaling pathway, and the absence of $IP_3$ leads to increased drought tolerance.

Phospholipase D (PLD) has also been involved in the signal transduction of water-deficit stress. PLD hydrolyzes phospholipids to generate phosphatidic acid (PA), another second messenger [23]. In guard cell, PLD activity mediates ABA-induced stomatal closure [24]. Drought and osmotic stress activate PLD and lead to transient increases in PA levels in plants [25-28]. PLD appears to be activated by osmotic stress through a G-protein [25]. Recently, it was demonstrated that PLDα and PA mediates salt stress signaling through the regulation of MAPK in Arabidopsis [29]. A specific PA species, which is increased by exposure to NaCl, bound to MPK6, and stimulated its activity. A plasma membrane $Na^+/H^+$ antiporter (SOS1) was identified as a downstream target of MPK6. These results suggest that PA plays a critical role in coupling MAPK cascades in response to salt stress [29].

## ROS

Water-deficit stress caused by drought or high salinity triggers the accumulation of ROS (superoxide, hydrogen peroxide, and hydroxyl radicals;

[30]. ROS are damaging agents that cause oxidative stress injury. Transgenic plants overexpressing ROS scavengers show increased tolerance to environmental stresses [30, 31]. However, ROS also act as signaling molecules that induce ROS scavengers and other protective mechanisms such as heat shock proteins (HSP), glutathione-S-transferases (GST), peroxidases, superoxide-dismutases and pathogenesis-related (PR) proteins, protecting the plant from damage [32].

The participation of ROS in water-deficit stress transduction is well-documented. ABA induces ROS production [33, 34], which are intermediate signals for ABA in mediating stomatal closure [34, 35]. Some genes related to osmotic stress signaling have been shown to be up-regulated by oxidative stress, including the transcription factor *DREB2A* [36]. The cellular mechanism by which ROS is produced in response to water-deficit stress remains unknown. Kwak *et al.* [37] have identified mutations in two of the ten NADPH oxidase (NOX) catalytic subunit genes in Arabidopsis (AtrbohD and AtrbohF) that abolish ABA-induced stomatal closure and ROS production, demonstrating that these two NOX are sources for ABA-triggered ROS production in guard cells.

The activation of oxidative stress signaling is highly conserved from yeast to mammals and involves MAPK cascades [38]. In Arabidopsis, it was reported that the MAPK AtMPK6, that is activated by osmotic stress is also activated by oxidative stress [39]. Therefore, it is likely that MAPKs that mediate osmotic stress signal transduction may also be used for ROS signaling in plants.

# PHOSPHOPROTEIN CASCADES

## Calcium Dependent Protein Kinases (CDPKs)

$Ca^{2+}$ is a key second messenger in water-deficit stress responses. Decoding complex signal-specific $Ca^{2+}$ signatures is accomplished by $Ca^{2+}$-binding proteins that function as $Ca^{2+}$ sensors, such as CDPKs. These proteins contain EF-hand motifs that bind $Ca^{2+}$ and a catalytic kinase domain whose activity is regulated by $Ca^{2+}$ binding. In *A. thaliana,* the CDPK family is one of the largest families and includes 34 unique CDPKs and CDPK-related protein kinases [40]. Several studies have shown that CDPKs have been implicated in signaling pathways in response to drought and salt stress [4-46].

Numerous studies indicate that $Ca^{2+}$ is likely to regulate gene expression [47] and it has been shown that some stress-induced $Ca^{2+}$ signals are perceived and transduced by CDPKs to activate stress-responsive gene expression. This was first demonstrated through transient expression experiments, in which an active AtCDPK1 induced the expression of the stress-responsive *HVA1* promoter–driven reporter gene in maize leaf protoplasts [43].

In *Mesembryanthemum crystallinum, McCPK1* expression increases in response to salt and osmotic stress in leaves. McCPK1 phosphorylates CSP1 (CDPK substrate protein 1) in the presence of $Ca^{2+}$ *in vitro*. CSP1 is constitutively located in the nucleus and acts as a transcription factor. McCPK1 is found in the plasma membrane in unstressed cells, and in the nucleus of cells exposed to salt stress and dehydration [45, 48]. These findings indicate that, under water-deficit stress, McCPK1 phosphorylates CSP1 in the nucleus, which changes the ability of CSP1 to regulate the expression of genes involved in the stress response. When stress is eliminated, McCPK1 returns to the plasma membrane. Thus, McCPK1 transmits the information from the plasma membrane into the nucleus to activate stress-inducible genes.

CDPKs also play important roles in ABA signaling. Arabidopsis CPK10 and CPK30 activate a stress and ABA inducible promoter [43]. Transcription levels of some CDPKs were shown to be stimulated by exogenous ABA [49, 50], their protein levels and activity are regulated by ABA as well [51]. Arabidopsis CPK32, CPK3, CPK6, CPK4, CPK11 and CPK10 were described as possible positive regulators [51-54], while CPK12 negatively regulates ABA signaling [55]. These observations suggest that different members of the CDPK family may constitute a complex regulation loop in ABA signaling, acting as positive or negative modulators. CDPKs are able to phosphorylate AREB/ABF (ABRE-binding proteins/ABRE-binding factors) transcription factors that mediate water-deficit stress responses by regulating the ABA-dependent gene expression [51, 55-58].

## SOS2 and SOS2-like Protein Kinases (PKSs)

As mentioned before, high salinity includes both an osmotic (physical) and an ionic (chemical) component. Salt stress disrupts plant ion homeostasis, resulting in toxic amounts of $Na^+$ in the cytoplasm and a deficiency of essential ions such as $K^+$. Different ion transporters function to limit $Na^+$ entry into and exit out of plant cells, to regulate $Na^+$ compartmentalization in the vacuole, and to selectively import $K^+$ over $Na^+$ into the cells. Genetic analysis of the *A. thaliana sos* mutants (salt-overlay-sensitive) identified the components SOS1, SOS2, and SOS3 involved in the signal transduction pathway activated by $Ca^{2+}$, that plays a key role in maintaining favorable ion ratios in the cytoplasm for salt stress tolerance

[59]. SOS3 is an EF hand-type $Ca^{2+}$-binding protein, able to sense specific salt stress-induced $Ca^{2+}$ signals [60]. SOS2 is a Ser/Thr kinase [61], and SOS1 is a $Na^+/H^+$ exchanger located on the plasma membrane [62, 63]. During salt stress conditions, the SOS2–SOS3 complex phosphorylates and activates the SOS1 antiporter [64, 65]. The *sos* mutants do not show altered responses to general osmotic stress or drought, indicating that SOS genes function specifically in the adaptive response to the ionic component of salt stress [66].

SOS3 is a $Ca^{2+}$ sensor essential for transducing the salt stress-induced $Ca^{2+}$ signal. SOS3 possesses three EF-hand motifs and binds $Ca^{2+}$ with low affinity compared to calmodulin [60]. The *sos3* mutation occurs in one of the EF-hand motifs and thus impairs the ability of the protein to bind $Ca^{2+}$. Myristoylation of SOS3 is required for its function in salt tolerance, since SOS3 regulate the activity of the membrane- bound ion transporter SOS1 [60]. SOS2 is a Ser/Thr protein kinase with a unique C-terminal regulatory domain and an N-terminal catalytic domain [61]. The regulatory region of SOS2 has an auto-inhibitory role that controls the protein kinase activity. The positive regulator SOS3 and the negative regulator type 2C protein phosphatase ABI2 bind to this regulatory region [67, 68]. The SOS3-binding domain is a 34-amino acid sequence known as the FISL motif, while the ABI2-binding sequence is known as the protein phosphatase interaction (PPI) motif. ABI2 dephosphorylates and thus deactivates SOS2 or SOS1 [69]. SOS1 is the only identified target of the SOS pathway up to date, although additional targets (e. g. vacuolar $Na^+/H^+$ antiporters [70]) have been proposed. SOS1 encodes a plasma membrane localized $Na^+/H^+$ antiporter [62], therefore, its function is to remove $Na^+$ from the cytoplasm and export it to the extracellular space.

The discovery of the $Ca^{2+}$ sensor SOS3 and its partner protein kinase SOS2 allowed the identification of eight *A. thaliana* genes coding for SOS3-like $Ca^{2+}$ sensor/binding proteins (SCaBPs) (also known as calcineurin B-like proteins, CBLs), and 23 SOS2-like protein kinases (PKSs) (also known as CBL-interacting protein kinases, CIPKs). Like SOS3, SCaBPs lack enzymatic activity and bind $Ca^{2+}$ through their EF hands. SCaBPs and PKSs are capable of numerous interactions *in vitro*. PKSs can be phosphorylated and this modification is important for the regulation of their activity. To the moment, PKS-phosphorylating kinases have not been reported, but a protein phosphatase 2C (PP2C) able to dephosphorylate PKS has been described [69]. PP2C interacts with PKS through the PKS conserved PPI motif within the C-terminal region [69].

The protein kinase CIPK1, interacts strongly with the calcium sensors CBL1 and CBL9 [71]. As reported for the CBL9 loss-of-function mutant, plants lacking either CBL1 or CIPK1 are hypersensitive to osmotic stress. In contrast to the *cbl1* mutant, but like the *cbl9* mutant, CIPK1 loss-of-function impairs ABA responsiveness. Therefore the CIPK1 kinase could be a converging point for ABA-dependent and ABA-independent stress responses

While considerable progress has been made in understanding the roles of SCaBP/CBLs and PKS/CIPKs, these studies are far from being completed and many questions are still open, mainly related to the functions of most members of these families and regarding the nature of their substrates.

## Mitogen Activated Protein Kinases (MAPKs)

In addition to the protein kinase pathways activated by $Ca^{2+}$, plants use other phosphoprotein modules for water-deficit stress signaling. In yeast, the HOG1 MAPK pathway is activated in response to hyperosmolarity. This pathway is responsible for the increased production of osmolytes allowing osmotic adjustment. It is possible that similar pathways exist in plants, since osmotic stress activates some components of the MAPK pathway. The Arabidopsis genome encodes approximately 60 MAPKKKs, 10 MAPKKs and 20 MAPKs. It has been reported that different MAPKs are activated at different times after the application of salt stress (e.g., [72]) and that different degrees of salt stress can activate distinct MAPKs [73].

The effect of reactive oxygen species (ROS) on the MAPK cascade results in the indirect activation of transcription factors. In Arabidopsis, the addition of $H_2O_2$ activates MAPK3 (AtMPK3) and AtMPK6 via activation of the MAPKKK ANP1 (NPK1-related protein kinase 1) [32] and strongly induces the expression of *AtNDPK2* (NUCLEOTIDE DIPHOSPHATE KINASE 2). Overexpression of *AtNDPK2* reduces the level of ROS, induces antioxidant genes and enhances tolerance to salt and oxidative stress. Since AtNDPK2 interacts with AtMPK3 and AtMPK6, and enhances the activity of AtMPK3 *in vitro*, it has been proposed that the ROS generated in response to water-deficit activate AtNDPK2, which in turn regulates the cellular redox state by the transient induction of antioxidant genes through the activation of the MAPK cascade [74].

## Protein Phosphatases

Phosphorylation/dephosphorylation of regulatory proteins serves as an "on-off" switch in the control of cellular activities. While the protein kinases and their functions have been known for many years, the understanding of the physiological roles of protein phosphatases, which are required partners of protein kinases in cellular control circuitries, has been developed more recently. Ser/Thr phosphatases are classified into Ser/Thr-specific protein phosphatases (PPP, which include subfamilies of PP1, PP2A, PP2B, PP4, PP5, PP6, PP7 and protein phosphatases with kelch-repeat domains) and the metal ion-dependent protein phosphatases (PPM, also known as PP2Cs). Members of Ser/Thr phosphatases of all subfamilies (except PP2B) have been identified in plants and different members of various subfamilies have been involved in water-deficit responses as positive or negative regulators [75].

Catalytic subunits of PP2A phosphatases occur either in association with a 65-kDa regulatory A subunit, or together with a third variable B-subunit in heterotrimeric complexes [75]. Evidence about the involvement of PP2A in water-deficit stress was found in several plant species [76]. In rice (*Oryza sativa*), all catalytic subunit genes (*OsPP2A-1-5*) are up-regulated in response to high salinity [77, 78]. In the same way, salt stress increases mRNA levels of potato PP2A catalytic subunits *StPP2Ac1*, *StPP2Ac2a*, *StPP2Ac2b* and *StPP2Ac3* [79]. Transgenic tobacco plants overexpressing the wheat (*Triticum aestivum*) *TaPP2Ac-1* exhibit enhanced drought tolerance [80], indicating that this PP2A catalytic subunit acts as positive regulator of salt stress adaptive response. Emerging evidence suggest that A and B subunits of PP2A are also involved in the regulation of water-deficit stress responses. Expression of alfalfa (*Medicago sativa*) *MsPP2A B$\beta$* subunit is induced by the phytohormone ABA, suggesting a function for this protein in the water-deficit stress response [81]. Arabidopsis loss-of-function mutants for *RCN1,* encoding a regulatory A subunit, exhibit increased sensitivity to osmotic stress, indicating that RCN1-regulated PP2A activity positively regulates the response to water-deficit [82].

While PP2As seem to positively regulate water-deficit stress response, PP2Cs are negative regulators of ABA signaling. The ABI1/ABI2 PP2Cs are central components in abscisic acid (ABA) signal transduction. Genetic analysis of *abi1* and *abi2* mutants, revertants and transgenic antisense plants showed that PP2Cs act as negative regulators of ABA signaling [83-86]. AtP2C-HAB1 (HYPERSENSITIVE TO ABA 1) is one of the closest relatives of ABI1 and ABI2. Overexpression of HAB1 impairs stomatal closure, diminishes ABA-induced gene expression, and leads to ABA insensitivity, indicating that HAB1 negatively regulates ABA signaling [87]. In addition to ABI1, ABI2 and HAB1, other three of the 76 known PP2Cs negatively regulate the ABA response in Arabidopsis: AtPP2CA/AHG3 (ABA HYPERSENSITIVE GERMINATION 3), HAB2 (HYPERSENSITIVE TO ABA 2), and AHG1 (ABA HYPERSENSITIVE GERMINATION 1) [88-90].

## Role of Abscisic Acid (ABA) in Water-Deficit Stress Responses

Upon stress, hormone production is increased in plants. Each hormone is mainly involved in some particular part of a stress response, but they may interact with one another in regulating stress signaling [5]. The plant hormone ABA has broad functions in plant growth and development, although it plays a major role in adaptation to water-deficit stress. This has been observed in ABA-deficient mutants of different plants such as Arabidopsis, tobacco, tomato or maize [91, 92], which can grow and develop relatively normal when no water stress is present. However, under drought or salt stresses, these mutants perform poorly or even die [59, 91, 93], showing that ABA is the central regulator of water-deficit stress resistance. ABA coordinates a complex regulatory network enabling plants to cope with decreased water availability [94-96].

Plant ABA content increases under drought or salt stress conditions, improving the plant capacity to cope with stress in different ways: water balance and cellular dehydration tolerance [59], which is done mainly by stimulating stomatal closure, changes in gene expression, and the accumulation of osmo-compatible solutes [94-97].

It is well known that induction of ABA biosynthesis is linked to changes in plant water status [98]. The turgor pressure was postulated to represent a critical factor of ABA accumulation some time ago [99]. Nevertheless, the signaling cascades that trigger ABA biosynthesis have not yet been characterized. Salt, drought, and to some extent cold stress, cause ABA accumulation. This ABA can be rapidly catabolized following the relief of the stress condition. ABA biosynthesis genes have been cloned [93, 100-102]. The expression of some of these genes was seen to be up-regulated by drought and/or by osmotic stress [93, 103-107], showing that the increased ABA levels under drought and salt stress are mainly achieved by induction of genes whose products are involved in the biosynthetic pathway (Figure 2). Interestingly, ABA1/ZEP, AAO and MSCU/ABA3 genes are all up-regulated by ABA [5, 93, 103], suggesting a positive feedback loop regulation and an important coordination controlled at multiple steps.

Numerous ABA signaling factors have been reported, including ABA receptors, protein kinases/phosphatases, transcription factors, RNA-binding proteins, chromatin remodeling factors and protein degradation enzymes [108, 109]. Although a core signaling model has been established, it is still necessary to answer many questions about the complexity of ABA signaling networks in plants to explain all ABA responses.

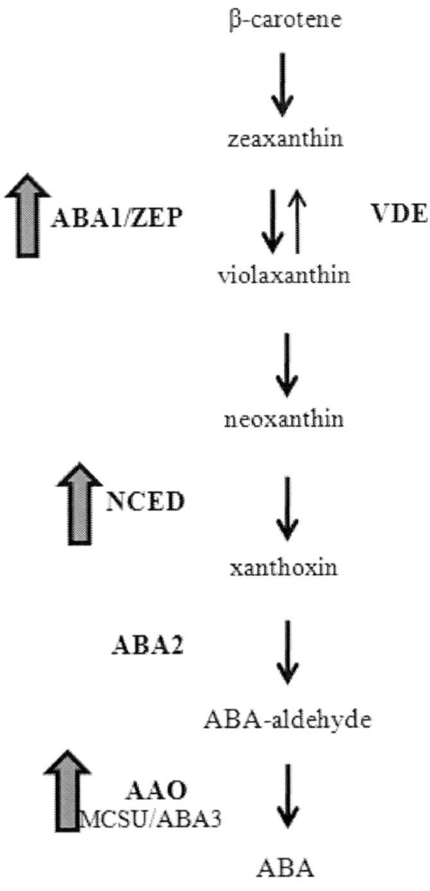


Figure 2. Pathway of ABA Biosynthesis. ABA is synthesized from β-carotene through the xanthophylls cycle. Zeaxanthin is interconverted into violaxanthin via epoxidation or de-epoxidation reactions catalyzed by zeaxanthin epoxidase (ABA1 or ZEP) and violaxanthin de-epoxidase (VDE), respectively. Violaxanthin is cleaved by 9-cis-epoxycarotenoid dioxygenase (NCED) to aldehyde xanthoxin. Xanthoxin is converted in a redox reaction catalyzed by the short-chain dehydrogenase/reductase ABA2 into abscisic aldehyde which is finally oxidized to ABA by abscisic aldehyde oxidase (AAO). Proper function of AAO requires the sulfurylated form of a molybdenum cofactor formed by the specific sulfurase MCSU/ABA3. Environmental stress such as drought, salt stress or sometimes cold stimulates the biosynthesis and accumulation of ABA by inducing genes coding for ABA biosynthetic enzymes (arrows indicated a positive regulation of expression of the enzyme).

## STRESS- AND ABA-REGULATED GENE EXPRESSION

Expression pattern of numerous plant genes has been reported to be regulated by drought and high salinity. Many of these genes are also regulated by ABA [109, 110]. High-throughput tools for gene expression analysis such as microarray technology has been used to discover many stress inducible genes involved in water-deficit stress response and classify them into two major groups. One group is mainly involved in water-deficit stress tolerance such as plasma membrane and vacuolar Na+/H+ antiporters responsible for ionic and osmotic homeostasis regulation by Na+ extrusion or compartmentation [111, 112], enzimes of osmolyte biosynthetic genes, LEA genes, detoxification enzymes, chaperones, proteases, ubiquitination-related enzymes and aquaporines [59]. These salt- and drought-induced genes appear to function in detoxification, damage prevention/limitation or repair [113]. The second set of genes regulated by water-deficit stress is involved in gene expression and signal transduction and includes protein kinases, transcription factors and enzymes involved in phospholipid metabolism.

*Cis*-acting elements and *trans*-acting factors are involved in gene expression regulation. Early-response genes are rapidly induced, and often transiently, since their induction does not require *de novo* protein synthesis. On the other hand, delayed-response genes are activated by stress more slowly and their expression is often sustained. The early response genes typically encode transcription factors that activate downstream delayed-response genes. The DREB/CBF (DRE-binding proteins/C-repeat-binding factor) and AREB/ABF (ABRE-binding proteins/ABRE-binding factors) gene families are examples of genes rapidly induced by either ABA, salt, drought or cold stress which encode transcription factors [59].

As mentioned before in this chapter, ABA mediates water-deficit stress responses, but is interesting to know whether all or most of the stress responses are dependent or not on ABA. ABA-deficient and ABA-insensitive mutants have allowed studying this question in the context of stress gene regulation. This kind of studies showed three sets of water-deficit stress responsive genes: completely independent of ABA, only partially ABA-dependent, and fully dependent on ABA [114]. There are two major *cis*-acting elements involved in water-deficit stress regulation of gene expression: ABRE and DRE that function in ABA-dependent and ABA-independent gene expression, respectively (Figure 3). However, genetic evidence suggests that stress-signaling pathways that were proposed as completely independent of ABA may not exist [2]. Table 1 summarizes the major *cis*-acting elements, DNA binding proteins (transcription factors), stress conditions which trigger changes in expression level, and examples of genes that present these elements in their promoters.

## ABA-DEPENDENT PATHWAY FOR STRESS REGULATION OF GENE EXPRESSION: ABRE CIS-ELEMENT AND AREB/ABF TRANSCRIPTION FACTORS

The ABA-dependent signaling pathway plays an essential role in stress-responsive gene expression during water-deficit stress. Promoter regions of several ABA-inducible genes revealed a conserved sequence, which was first identified as a *cis*-acting element named ABA responsive element (ABRE). This is the most important *cis*-acting element involved in ABA-dependent regulation of gene expression. Deletion and mutational analysis of promoters shows that a single copy of ABRE is not sufficient for ABA-responsive transcription. ABRE needs coupling elements, at least one, to constitute an ABA-responsive complex. And even another copy of ABRE might function as coupling element.

Yeast one-hybrid screening method allowed the isolation of transcription factors that can bind ABRE elements, called ABRE-binding proteins (AREB) or ABRE-binding factors (ABFs). The ABA-dependent pathway activates the transcription factors AREB/ABF to induce stress-responsive genes with ABRE elements in their promoters (Figure 3).

It has been demonstrated that AREB/ABFs must be activated in an ABA-dependent manner to induce the transcription of target genes. This ABA-mediated signal is probably a phosphorylation of the AREB/ABF proteins [110, 115]. The first kinases reported to be involved in the ABA-dependent activation of AFBs were the SnRK2s (SNF1-related protein kinase subfamily 2; [115-117]). Several lines of evidence indicate that SnRK2 are involved in ABA and/or stress signaling pathways. Overexpression of SnRK2C improves Arabidopsis drought tolerance [118]. All ten SnRK2 members are induced by hiperosmotic stress, and some of them by ABA [119]. It has been reported the phosphorylation of AREB/ABFs by different SnRK2s [116, 120, 121].

In recent years, calcium-dependent protein kinases (CDPKs) have also been identified as regulators of AREB/ABF transcription factors. As mentioned before, CDPKs are central components in ABA signaling, acting as positive or negative regulators. It has been shown that CDPKs phosphorylates AREB/ABF proteins. AtCPK32 was isolated by a yeast two-hybrid screen as an ABF4- interacting

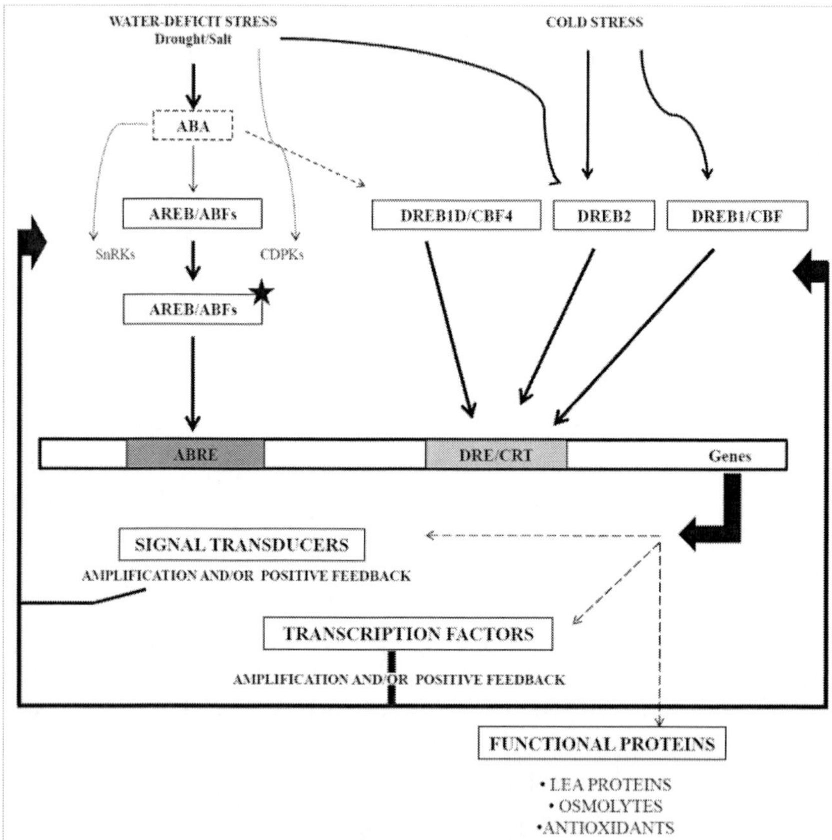


Figure 3. Transcriptional regulation of stress-responsive genes in response to water-deficit stress. Water-deficit stress induces ABA accumulation which in turn activates ABA-dependent signaling pathways. Water-deficit stress also triggers ABA-independent pathways trough DREB2 proteins. ABA-dependent signaling involves AREB/ABFs phosphorylation by SnRKs and/or CDPKs. Activated AREB/ABFs bind to ABRE elements and induce transcription of stress-responsive genes. DREB transcription factors bind to the DRE sequence. DREB1D is involved in ABA-dependent signaling, while DREB2 mediates water-deficit stress response through the ABA–independent pathway. Both ABA dependent and ABA-independent signaling pathways appear to be involved in water-deficit stress tolerance. Cold stress signaling involves DREB2 and DREB1/CBFs proteins. There are multiple stress perception and signaling pathways, some of them are specific, but others may cross-talk at various steps, increasing the complexity of the network. Changes in gene expression involve signal transducers and transcription factors that amplify the stress signal, and functional proteins that contribute to protect the cell from the damage caused by the stress.

protein. AtCPK32 can phosphorylate ABF4 *in vitro*, and its overexpression affects both ABA sensitivity and the expression of a number of ABF4-regulated genes [52]. Both Arabidopsis CPK4 and CPK11 phosphorylate ABF1 and ABF4 *in vitro*. Loss-of-function mutations of *CPK4* and *CPK11* resulted in ABA-insensitive phenotypes and decreased tolerance of seedlings to salt stress, suggesting that the two kinases may positively regulate ABA signaling through ABF1 and ABF4 tanscription factors [51]. Recently, it has been shown that Arabidospis CPK12, which negatively regulates early events of ABA signaling, phosphorylates both ABF1 and ABF4 *in vitro* [55].

Not all the ABA-dependent stress inducible genes present ABRE motifs in their promoter regions. MYC and MYB recognition sites are *cis*-acting regulatory elements involved in ABA dependent gene expression. These two elements are recognized by MYC-like and MYB transcription factors respectively. These transcription factors are synthesized in response to endogenous ABA accumulation, corresponding with a late stage of the stress response. Overexpression of these genes leads to osmotic stress tolerance [122].

## ABA-Independent Stress Responsive Gene Expression: DRE Cis-Element and CBF/DREB1 Transcription Factors

ABA-deficient mutants have shown drought and cold induced gene expression despite the low level or even absence of the hormone, which indicates an alternative ABA-independent regulation. ABA-independent pathway of stress responsive gene expression has been described through DRE *cis*-element regulation. DRE, named for dehydration responsive element, is an essential *cis*-acting element of nine conserved base pair that is sufficient, in a single copy form, to regulate the expression of *RD29A* gene. Therefore, it does not require extra elements for its function. Other similar *cis*-acting elements, called C-repeat (CRT) and low-temperature-responsive element (LTRE), containing the core DRE sequence, have also been reported in cold regulated promoters [123, 124].

Yeast one-hybrid screenings allowed the isolation of transcription factors from AP2 family that bind DRE elements present in the promoter of stress responsive genes, such as C-repeat-binding factor 1 (CBF1), DRE-binding protein 1A (DREB1A) and DREB2A [125, 133]. Now, many homologues to CBF1 and DREB1A and DREB2A have been described [125-127]. Both DREB1/CBF and DREB2 proteins bind to DRE, but DREB1/CBFs have been related to cold responsive gene expression, whereas DREB2s are involved in water deficit-responsive gene expression. In this way, expression of DREB1/CBF genes is induced by cold stress, while DREB2A and DREB2B are activated by water-deficit stress and activates induction of target stress-responsive genes [125]. Initially, this pathway was described as ABA-independent [125], however more recent research showed that gene expression of CBF4, a DREB1-related transcription factor, is up-regulated by drought and ABA, but not by cold stress, and its over-expression resulted in constitutive expression of DRE-containing stress responsive genes, as well as enhanced tolerance to drought and cold stress [128]. These findings show that regulation of DRE elements might also be mediated by ABA, and reveal a cross-talk between drought- and cold-responsive gene expression. Moreover, Narusaka *et al.* [129] showed that DRE and ABRE elements are not isolated regulatory sequences inside a promoter region but are interdependent in the expression of the *RD29A* gene.

**Table 1. Major *cis*-acting elements and DNA binding proteins involved in transcription regulation under water-deficit stress**

| *Cis*-element | Gene example | Transcription factors | Family | Stress condition involved | References |
|---|---|---|---|---|---|
| DRE | rd29A | DREB2A (*A. thaliana*)<br>DREB2B (*A. thaliana*)<br>CBF1/DREB1B (*A. thaliana*)<br>CBF2/DREB1C (*A. thaliana*)<br>CBF3/DREB1A (*A. thaliana*) | AP2 | Dehydratation<br>Salt<br>ABA<br>Cold | [125, 126, 127, 132, 133] |
| ABRE | Em<br>Osem<br>rd29b<br>rd29A<br>SILAP | EmBP1 (*T. aestivum*)<br>TAF-1 (*N. tabcum*)<br>OSBZ8 (*O. sativa*)<br>OsZIP-1a (*O. sativa*)<br>TRAB1 (*O. sativa*)<br>AREB1/ABF2 (*A. thaliana*)<br>AREB2/ABF4 (*A. thaliana*)<br>AREB3/ABF1 (*A. thaliana*)<br>ABF3 (*A. thaliana*)<br>SlAREB/SlAREB1 (*S. lycopersicum*)<br>SlAREB2 (*S. lycopersicum*) | b-ZIP | ABA<br>Dehydratation<br>Salt<br>Cold | [56, 115, 134, 135, 136, 137, 138, 139, 140, 141, 142] |

New insights into transcriptional cascades in stress gene regulation open up new possibilities in molecular engineering field, in which generation of plants with increased stress tolerance are being designed. Because of their nature as master switches in regulatory networks, transcription factors of DREB/CBF or AREB/ABF family have emerged as a new tool for engineering abiotic stress tolerance, since many genes involved in stress response can be simultaneously regulated by a single gene [130, 131].

## PHYSIOLOGICAL AND BIOCHEMICAL RESPONSES TO WATER-DEFICIT STRESS

Water-deficit stress induces physiological and biochemical changes at the cellular level such as turgor loss, changes in membrane fluidity and composition and protein-protein interactions [143]. Plants can maintain turgor during water-deficit by accumulating osmolytes, compatible solutes or osmoprotectants, which act as osmotic balancing agents. These molecules are synthesized in plant cells in response to water-deficit stress and are degraded after stress relief [144]. Osmolytes include amino acids such as proline, polyhydric alcohols (pinitol, mannitol), and quaternary ammonium compounds (glycine betaine, proline betaine, choline-O-sulphate). Synthesis and accumulation of osmolytes is widespread in plants, however the distribution of specific compatible solutes varies among plant species [145]. Osmolytes are highly soluble and can accumulate at high concentrations in the cytoplasm without interfering with cellular metabolism, equilibrating osmotic potential to avoid loss of turgor. It has been proposed that under water-deficit conditions, the hydroxyl group of polyhydroxy compounds can form a hydrogen bond with the polar heads of membrane phospholipids, and that these hydrophobic interactions are important for membrane stability [146-148].

The *LEA* (*late embryogenesis abundant*) genes were first identified as genes induced in seeds during maturation and desiccation. Some LEA genes expression is induced in the vegetative tissues of plants exposed to water-deficit stress. Although there is a strong association of LEA proteins with water-deficit stress tolerance, their function remains poorly understood [149]. Most LEA proteins are highly hydrophilic, which is consistent with their cytoplasmic location. Based on this observation, it has been suggested that they might act as hydration buffers, slowing down the rate of water loss during dehydration. LEA proteins might also serve as space fillers to prevent cellular collapse at low water activities [150]. A number of studies have shown that they can protect proteins such as lactate dehydrogenase, malate dehydrogenase, fumarase and citrate synthase against desiccation, due to their ability to prevent protein aggregation. During desiccation, LEA proteins may behave as molecular shields exerting an excluded volume effect that could serve to decrease interaction between partially denatured polypeptides with the potential to aggregate [151]. It has also been proposed that LEA proteins act as antioxidants and as membrane stabilizers during water stress, either by direct interaction or by acting as molecular shields [151].

## Conclusion

The understanding of water-deficit stress responses has taken a big step forward in the last decade, however, it is evident that the subject is very complex and the comprehension of the whole-plant stress response mechanism is still very limited. One of the problems is that the primary sensors of water-deficit stress signals are as yet unidentified, and how these sensors transmit information to protein kinases remains unknown. Another question is which are the substrates of the protein kinases activated by water-deficit stress, and how are these kinases connected with transcription factors.

Many genes whose expression changes in response to water-deficit stress have been identified. The challenge to be met is to determine their regulons, identify the transcription factors controlling them, and elucidate the physiological roles of the proteins encoded by the regulons in the development of stress tolerance. Forward and reverse genetics approaches will be useful to dissect these complex pathways. Spatial and temporal expression profiling, biochemical characterization of the signaling components and identification of signaling complexes will allow the determination of cross-talk and specificity of the signaling pathways activated in response to water-deficit stress.

## References

[1] Chinnusamy, V.; Schumaker, K. & Zhu, J.K. Molecular genetic perspectives on cross-talk and specificity in abiotic stress signalling in plants. *Journal of Experimental Botany*, 2004, 55, 225-236.

[2] Mahajan, S. & Tuteja, N. Cold, salinity and drought stresses: an overview. *Archives of Biochemistry and Biophysics*, 2005, 444, 139-58.

[3] Xu, Z.; Zhou, G. & Shimizu, H. Plant responses to drought and rewatering. *Plant Signaling and Behavior*, 2010, 5, 649-654.

[4] Hirayama, T. & Shinozaki, K. Research on plant abiotic stress responses in the post-genome era: past, present and future. *The Plant Journal*, 2010, 61, 1041-1052.

[5] Xiong, L.; Schumaker, K.S. & Zhu, J.K. Cell signaling during cold, drought, and salt stress. *The Plant Cell*, 2002, 14, S165-183.

[6] Suzuki, I.; Dmitry, A.L.; Kanesaki, Y.; Mikami, K. & Murata, N. The pathway for perception and transduction of low-temperature signals in Synechocystis. *The EMBO Journal*, 2000, 19, 1327-1334.

[7] Aguilar, P.S.; Hernandez-Arriaga, A.M.; Cybulski, L.E.; Erazo, A.C. & de Mendoza, D. Molecular basis of thermosensing: a two component signal transduction thermometer in Bacillus subtilis. *The EMBO Journal*, 2001, 20, 1681-1691.

[8] Maeda, T.; Wurgler-Murphy, S.M. & Saito, H. A two-component system that regulates an osmosensing MAP kinase cascade in yeast. *Nature*, 1994, 369, 242-245.

[9] Urao, T.; Yakubov, B.; Satoh, R.; Yamaguchi-Shinozaki, K.; Seki, M.; Hirayama, T. & Shinozaki, K. A transmembrane hybrid-type histidine kinase in Arabidopsis functions as an osmosensor. *The Plant Cell*, 1999, 11, 1743-1754.

[10] Tamura, T.; Hara, K.; Yamaguchi, Y.; Koizumi, N. & Sano, H. Osmotic stress tolerance of transgenic tobacco expressing a gene encoding a membrane-located receptor-like protein from tobacco plants. *Plant Physiology*, 2003, 131, 454-462.

[11] Nakamura, K. & Sano, H. A plasma-membrane linker for the phosphoinositide-specific phospholipase C in tobacco plants. *Plant Signaling and Behavior*, 2009, 4, 26-9.

[12] Meijer, H.J. & Munnik, T. Phospholipid-based signaling in plants. *Annual Review of Plant Biology*, 2003, 54, 265-306.

[13] Knight, H. Calcium signaling during abiotic stress in plants. *International Review of Cytology*, 2000, 195, 269–325.

[14] Sanders, D.; Brownlee, C. & Harper, J.F. Communicating with calcium. *The Plant Cell*, 1999, 11, 691–706.

[15] Schroeder, J.I.; Allen, G.J.; Hugouvieux, V.; Kwak, J.M. & Waner, D. Guard cell signal transduction. *Annual Review of Plant Physiology and Plant Molecular Biology*, 2001, 52, 627–658.

[16] Pical, C.; Westergren, T.; Dove, S.K.; Larsson, C. & Sommarin, M. Salinity and hyperosmotic stress induce rapid increases in phosphatidylinositol 4,5-bisphosphate, diacylglycerol pyrophosphate, and phosphatidylcholine in *Arabidopsis thaliana* cells. *Journal of Biological Chemistry*, 1999, 274, 38232–38240.

[17] DeWald, D.B.; Torabinejad, J.; Jones, C.A.; Shope, J.C.; Cangelosi, A.R.; Thompson, J.E.; Prestwich, G.D. & Hama, H. Rapid accumulation of

phosphatidylinositol 4,5-bisphosphate and inositol 1,4,5-trisphosphate correlates with calcium mobilization in salt-stressed Arabidopsis. *Plant Physiology*, 2001, 126, 759–769.

[18] Mikami, K.; Katagiri, T.; Luchi, S.; Yamaguchi-Shinozaki, K. & Shinozaki, K. A gene encoding phosphatidylinositol 4-phosphate 5-kinase is induced by water stress and abscisic acid in *Arabidopsis thaliana. The Plant Journal*, 1998, 15, 563–568.

[19] Hirayama, T.; Ohto, C.; Mizoguchi, T. & Shinozaki, K. A gene encoding a phosphatidylinositol-specific phospholipase C is induced by dehydration and salt stress in *Arabidopsis thaliana. Proceedings of the National Academy of Sciences of the United States of America*, 1995, 92, 3903–3907.

[20] Kopka, J.; Pical, C.; Gray, J.E. & Muller-Rober, B. Molecular and enzymatic characterization of three phosphoinositide- specific phospholipase C isoforms from potato. *Plant Physiology*, 1998, 116, 239–250.

[21] Xiong, L.; Lee, B.H.; Ishitani, M.; Lee, H.; Zhang, C. & Zhu, J.K. *FIERY1* encoding an inositol polyphosphate 1-phosphatase is a negative regulator of abscisic acid and stress signaling in Arabidopsis. *Genes and Development*, 2001, 15, 1971–1984.

[22] Perera, I.Y.; Hung, C.Y.; Moore, C.D.; Stevenson-Paulik, J. & Boss, W.F. Transgenic Arabidopsis plants expressing the type 1 inositol 5-phosphatase exhibit increased drought tolerance and altered abscisic acid signaling. *The Plant Cell*, 2008, 20, 2876-2893.

[23] Wang, X. The role of phospholipase D in signaling cascades. *Plant Physiology*, 1999, 120, 645–651.

[24] Jacob, T.; Ritchie, S.; Assmann, S.M. & Gilroy, S. Abscisic acid signal transduction in guard cells is mediated by phospholipase D activity. *Proceedings of the National Academy of Sciences of the United States of America*, 1999, 9, 12192–12197.

[25] Frank, W.; Munnik, T.; Kerkmann, K.; Salamini, F. & Bartels, D. Water-deficit triggers phospholipase D activity in the resurrection plant Craterostigma plantagineum. *The Plant Cell*, 2000, 12, 111–123.

[26] Munnik, T.; Meijer, H.J.G.; ter Riet, B.; Frank, W.; Bartels, D. & Musgrave, A. Hyperosmotic stress stimulates phospholipase D activity and elevates the levels of phosphatidic acid and diacylglycerol pyrophosphate. *The Plant Journal*, 2000, 22, 147–154.

[27] Katagiri, T.; Takahashi, S. & Shinozaki, K. Involvement of a novel Arabidopsis phospholipase D, AtPLDdelta, in dehydration inducible

accumulation of phosphatidic acid in stress signaling. *The Plant Journal*, 2001, 26, 595–605.

[28] Hong, Y.; Zhang, W. & Wang, X. Phospholipase D and phosphatidic acid signalling in plant response to drought and salinity. *Plant Cell and Environment*, 2010, 33, 627-635.

[29] Yu, L.; Nie, J.; Cao, C.; Jin, Y.; Yan, M.; Wang, F.; Liu, J.; Xiao, Y.; Liang, Y. & Zhang, W. Phosphatidic acid mediates salt stress response by regulation of MPK6 in Arabidopsis thaliana. *New Phytologist*, 2010, 188, 762-773.

[30] Hasegawa, P.M.; Bressan, R.A.; Zhu, J.K. & Bohnert, H.J. Plant cellular and molecular responses to high salinity. *Annual Review of Plant Physiology and Plant Molecular Biology*, 2000, 51, 463–499.

[31] Kocsy, G.; Galiba, G. & Brunold, C. Role of glutathione in adaptation and signaling during chilling and cold acclimation in plants. *Physiologia Plantarum*, 2001, 113, 158–164.

[32] Kovtun, Y.; Chiu, W.L.; Tena, G. & Sheen, J. Functional analysis of oxidative stress-activated mitogen-activated protein kinase cascade in plants. *Proceedings of the National Academy of Sciences of the United States of America*, 97, 2000, 2940–2945.

[33] Guan, L.M.; Zhao, J. & Scadalios, J.G. *Cis*-elements and trans-factors that regulate expression of the maize *Cat1* antioxidant gene in response to ABA and osmotic stress: $H_2O_2$ is the likely intermediary signaling molecule for the response. *The Plant Journal*, 2000, 22, 87–95.

[34] Pei, Z.M.; Murata, Y.; Benning, G.; Thomine, S.; Klusener, B.; Allen, G.J.; Grill, E. & Schroeder, J.I. Calcium channels activated by hydrogen peroxide mediate abscisic acid signaling in guard cells. *Nature*, 2000, 406, 731–734.

[35] Zhang, X.; Zhang, L.; Dong, F.; Gao, J.; Galbraith, D.W. & Song, C.P. Hydrogen peroxide is involved in abscisic acid-induced stomatal closure in *Vicia faba*. *Plant Physiology*, 2001, 126, 1438–1448.

[36] Desikan, R.; Mackerness, S.A.H.; Hancock, J.T. & Neill, S.J. Regulation of the Arabidopsis transcriptome by oxidative stress. *Plant Physiology*, 2001, 127, 159–172.

[37] Kwak, J.M.; Mori, I.C.; Pei, Z.M.; Leonhardt, N.; Torres, M.A.; Dangl, J.L.; Bloom, R.E.; Bodde, S.; Jones, J.D. & Schroeder, J.I. NADPH oxidase AtrbohD and AtrbohF genes function in ROS-dependent ABA signaling in Arabidopsis. *The EMBO Journal*, 2003, 22, 2623–2633.

[38] Xiong, L. & Zhu, J.K. Molecular and genetic aspects of plant responses to osmotic stress. *Plant Cell and Environment*, 2002, 25, 131–139.

[39] Yuasa, T.; Ichimura, K.; Mizoguchi, T. & Shinozaki, K. Oxidative stress activates ATMPK6, an Arabidopsis homologue of MAP kinase. *Plant and Cell Physiology*, 2001, 42, 1012–1016.

[40] Harmon, A.C. Calcium-regulated protein kinases of plants. *Gravitational and Space Biology Bulletin*, 2003, 16, 83–90.

[41] Urao, T.; Katagiri, T.; Mizoguchi, T.; Yamaguchi-Shinozaki, K.; Hayashida, N. & Shinozaki, K. Two genes that encode Ca2_-dependent protein kinases are induced by drought and high salt stresses in *Arabidopsis thaliana. Molecular and General Genetics*, 1994, 224, 331–340.

[42] Pei, Z.M.; Ward, J.M.; Harper, J.F. & Schroeder, J.I. A novel chloride channel in Vicia faba guard cell vacuoles activated by the serine/threonine kinase, CDPK. *The EMBO Journal*, 1996, 15, 6564–6574.

[43] Sheen, J. $Ca^{2+}$-dependent protein kinases and stress signal transduction in plants. *Science*, 1996, 274, 1900–1902.

[44] Saijo, Y.; Hata, S.; Kyozuka, J.; Shimamoto, K. & Izui, K. Over-expression of a single $Ca^{2+}$-dependent protein kinase confers both cold and salt/drought tolerance on rice plants. *The Plant Journal*, 2000, 23, 319–327.

[45] Patharkar, O.R., & Cushman, J.C. A stress-induced calcium- dependent protein kinase from *Mesembryanthemum crystallium* phosphorylates a two-component pseudo-response regulator. *The Plant Journal*, 2000, 24, 679–691.

[46] Capiati, D.A.; País, S.M. & Téllez-Iñón, M.T. Wounding increases salt tolerance in tomato plants: evidence on the participation of calmodulin-like activities in cross-tolerance signalling. *Journal of Experimental Botany*, 2006.57, 2391-2400.

[47] Reddy, A.S.N.; Ali, G.S.; Celesnik, H. & Day, I.S. Coping with Stresses: Roles of Calcium- and Calcium/Calmodulin-Regulated Gene Expression. *The Plant Cell*, 2011, 23, 2010-2032.

[48] Chehab, E.W.; Patharker, O.R.; Hegeman, A.D.; Taybi, T. & Cushman, J.C. Autophosphorylation and subcellular localization dynamics of a salt- and water deficit induced calcium-dependent protein kinase from ice plant. *Plant Physiology*, 2004, 135, 1430–1446.

[49] Li, Z. & Komatsu, S. Molecular cloning and characterization of calreticulin, a calcium-binding protein involved in the regeneration of rice cultured suspension cells. European Journal of Biochemistry, 2000, 267, 737-745.

[50] Yoon, G.M.; Cho, H.S.; Ha, H.J.; Liu, J.R. & Lee, H.S. Characterization of NtCDPK1, a calcium-dependent protein kinase gene in *Nicotiana tabacum*, and the activity of its encoded protein. *Plant Molecular Biology*, 1999, 39, 1991-1001.

[51] Zhu, S.Y.; Yu, X.C.; Wang, X.J.; Zhao, R.; Li, Y.; Fan, R.C.; Shang, Y.; Du, S.Y.; Wang X.F.; Wu, F.Q.; Xu, Y.H.; Zhang, X.Y. & Zhang, D.P. Two calcium-dependent protein kinases, CPK4 and CPK11, regulate abscisic acid signal transduction in Arabidopsis. *The Plant Cell*, 2007, 19, 3019-3036.

[52] Choi, H.I.; Park, H.J.; Park, J.H.; Kim, S.; Im, M.Y.; Seo, H.H.; Kim, Y.W.; Hwang, I. & Kim, S.Y. Arabidopsis calcium-dependent protein kinase AtCPK32 interacts with ABF4, a transcriptional regulator of abscisic acid-responsive gene expression, and modulates its activity. *Plant Physiology*, 2005, 139, 1750-1761.

[53] Mori, I.C.; Murata, Y.; Yang, Y.; Munemasa, S.; Wang, Y.F.; Andreoli, S.; Tiriac, H.; Alonso, J.M.; Harper, J.F.; Ecker, J.R.; Kwak, J.M. & Schroeder, J.I. CDPKs CPK6 and CPK3 function in ABA regulation of guard cell S-type anion- and $Ca^{2+}$-permeable channels and stomatal closure. *PLoS Biology*, 2006, 4:e327.

[54] Zou, J.J.; Wei, F.J.; Wang, C.; Wu, J.J.; Ratnasekera, D.; Liu, W.X. & Wu, W.H. Arabidopsis calcium-dependent protein kinase CPK10 functions in abscisic acid- and $Ca^{2+}$-mediated stomatal regulation in response to drought stress. *Plant Physiology*, 2010, 154, 1232-1243.

[55] Zhao, R.; Sun, H.L.; Mei, C.; Wang, X.J.; Yan, L.; Liu, R.; Zhang, X.F.; Wang, X.F. & Zhang, D.P. The Arabidopsis $Ca^{2+}$-dependent protein kinase CPK12 negatively regulates abscisic acid signaling in seed germination and post-germination growth. *New Phytologist*, 2011, 192, 61-73.

[56] Choi, H.; Hong, J.; Ha, J.; Kang, J. & Kim, S.Y. ABFs, a family of ABA-responsive element binding factors. *Journal of Biological Chemistry*, 2000, 275, 1723-1730.

[57] Uno, Y.; Furihata, T.; Abe, H.; Yoshida, R.; Shinozaki, K. & Yamaguchi-Shinozaki, K. Arabidopsis basic leucine zipper transcription factors involved in an abscisic acid-dependent signal transduction pathway under drought and high-salinity conditions. *Proceedings of the National Academy of Sciences of the United States of America*, 2000, 97, 11632-11637.

[58] Choi, H.I., Park, H.J., Park, J.H., Kim, S., Im, M.Y., Seo, H.H., Kim, Y.W., Hwang, I. & Kim, S.Y. Arabidopsis calcium-dependent protein kinase AtCPK32 interacts with ABF4, a transcriptional regulator of abscisic acid-responsive gene expression, and modulates its activity. *Plant Physiology*, 2005, 139, 1750-1761.

[59] Zhu, J.K. Salt and drought stress signal transduction in plants. *Annual Review of Plant Biology*, 2002, 53, 247–273.

[60] Ishitani, M.; Liu, J.; Halfter, U.; Kim, C.S.; Shi, W. & Zhu, J.K. SOS3 function in plant salt tolerance requires N-myristoylation and calcium-binding. *The Plant Cell*, 2000, 12, 1667–1677.

[61] Liu, J.; Ishitani, M.; Halfter, U.; Kim, C.S. & Zhu, J.K. The Arabidopsis thaliana SOS2 gene encodes a protein kinase that is required for salt tolerance. *Proceedings of the National Academy of Sciences of the United States of America*, 2000, 97, 3730–3734.

[62] Shi, H.; Ishitani, M.; Kim, C.S. & Zhu, J.K. The Arabidopsis thaliana salt tolerance gene SOS1 encodes a putative Na+/H+ antiporter. *Proceedings of the National Academy of Sciences of the United States of America*, 2000, 97, 6896–6901.

[63] Qiu, Q.S.; Guo, Y.; Dietrich, M.A.; Schumaker, K.S. & Zhu, J.K. Regulation of SOS1, a plasma membrane $Na^+/H^+$ exchanger in Arabidopsis thaliana, by SOS2 and SOS3. *Proceedings of the National Academy of Sciences of the United States of America*, 2002, 99, 8436–8441.

[64] Quintero, F.J.; Ohta, M.; Shi, H.; Zhu, J.K. & Pardo, J.M. Reconstitution in yeast of the Arabidopsis SOS signaling pathway for $Na^+$ homeostasis. *Proceedings of the National Academy of Sciences of the United States of America*, 2002, 99, 9061–9066.

[65] Batelli, G.; Verslues, P.E.; Agius, F.; Qiu, Q.; Fujii, H.; Pan, S.; Schumaker, K.S., Grillo, S; & Zhu, J.K. SOS2 promotes salt tolerance in part by interacting with the vaculoar H?-ATPase and upregulating its transport activity. *Molecular and Cellular Biology*, 2007, 27, 7781–7790.

[66] Zhu, J.K. Cell signaling under salt, water and cold stresses. *Plant Biology*, 2001, 4:401–406

[67] Guo, Y.; Halfter, U.; Ishitani, M. & Zhu, J.K. Molecular characterization of functional domains in the protein kinase SOS2 that is required for plant salt tolerance. *The Plant Cell*, 2001, 13, 1383–1400.

[68] Sanchez-Barrena, M.J.; Fujii, H.; Angulo, I.; Martinez-Ripoll, M.; Zhu, J.K. & Albert, A. The structure of the C-terminal domain of the protein kinase AtSOS2 bound to the calcium sensor AtSOS3. *Molecular Cell*, 2007 26, 427–435.

[69] Ohta, M.; Guo, Y.; Halfter, U. & Zhu, J.K. A novel domain in the protein kinase SOS2 mediates interaction with the protein phosphatase 2C ABI2. *Proceedings of the National Academy of Sciences of the United States of America*, 2003, 100, 11771–11776.

[70] Qiu, Q.S.; Guo, Y.; Quintero, F.J.; Pardo, J.M.; Schumaker, K.S. & Zhu, J.K. Regulation of vacuolar Na+/H+ exchange in Arabidopsis thaliana by the SOS pathway. *Journal of Biological Chemistry*, 2004, 279, 207-215.

[71] D'Angelo, C.; Weinl, S.; Batistic, O.; Pandey, G.K.; Cheong, Y.H.; Schültke, S.; Albrecht, V.; Ehlert, B.; Schulz, B.; Harter, K.; Luan, S.; Bock, R. & Kudla, J. Alternative complex formation of the Ca-regulated protein kinase CIPK1 controls abscisic acid-dependent and independent stress responses in Arabidopsis. *The Plant Journal,* 2006, 48, 857-872.
[72] Mikolajczyk, M.; Olubunmi, S.A.; Muszynska, G.; Klessig, D.F. & Dobrowolska, G. Osmotic stress induces rapid activation of a salicylic acid-induced protein kinase and a homolog of protein kinase ASK1 in tobacco cells. *The Plant Cell*, 2000, 12, 165–178.
[73] Munnik, T.; Ligterink, W.; Meskiene, I.; Calderini, O.; Beyerly, J.; Musgrave, A. & Hirt, H. Distinct osmo-sensing protein kinase pathways are involved in signalling moderate and severe hyper-osmotic stress. *The Plant Journal*, 1999, 20, 381–388.
[74] Laloi, C.; Apel, K. & Danon, A. Reactive oxygen signalling: the latest news. *Current Opinion in Plant Biology*, 2004, 7, 323–328.
[75] Farkas, I.; Dombrádi, V.; Miskei, M.; Szabados, L. & Koncz, C. Arabidopsis PPP family of serine/threonine phosphatases. *Trends in Plant Science*, 2007; 12, 169-176.
[76] País, S.M.; Téllez-Iñón, M.T. & Capiati, D.A. Serine/threonine protein phosphatases type 2A and their roles in stress signaling. *Plant Signaling and Behavior*, 2009, 4, 1013-1015.
[77] Yu, R.M.; Zhou, Y.; Xu, Z.F.; Chye, M.L. & Kong, R.Y. Two genes encoding protein phosphatase 2A catalytic subunits are differentially expressed in rice. *Plant Molecular Biology*, 2003, 51, 295-311.
[78] Yu, R.M.; Wong, M.M.; Jack, R.W. & Kong, R.Y. Structure, evolution and expression of a second subfamily of protein phosphatase 2A catalytic subunit genes in the rice plant (*Oryza sativa* L.). *Planta*, 2005; 222, 757-768.
[79] País, S.M.; González, M.A.; Téllez-Iñón, M.T. & Capiati, D.A. Characterization of potato (*Solanum tuberosum*) and tomato (*Solanum lycopersicum*) protein phosphatases type 2A catalytic subunits and their involvement in stress responses. *Planta*, 2009; 230, 13-25.
[80] Xu, C.; Jing, R.; Mao, X.; Jia, X. & Chang, X. A wheat (*Triticum aestivum*) protein phosphatase 2A catalytic subunit gene provides enhanced drought tolerance in tobacco. *Annals of Botany*, 2007; 99, 439-450.
[81] Tóth, E.C.; Vissi, E.; Kovács, I.; Szöke, A.; Ariño, J.; Gergely, P.; Dudits, D. & Dombrádi, V. Protein phosphatase 2A holoenzyme and its subunits from *Medicago sativa*. Plant Molecular Biology, 2000; 43, 527-536.

[82] Blakeslee, J.J.; Zhou, H.W.; Heath, J.T.; Skottke, K.R.; Barrios, J.A.; Liu S.Y. & DeLong, A. Specificity of RCN1-mediated protein phosphatase 2A regulation in meristem organization and stress response in roots. *Plant Physiology*, 2008, 146, 539-553.

[83] Merlot, S.; Gosti, F.; Guerrier, D.; Vavasseur, A. & Giraudat, J. The ABI1 and ABI2 protein phosphatases 2C act in a negative feedback regulatory loop of the abscisic acid signaling pathway. *The Plant Journal*, 2001, 25, 295–303.

[84] Gosti, F.; Beaudoin, N.; Serizet, C.; Webb, A.A.; Vartanian, N. & Giraudat, J. ABI1 protein phosphatase 2C is a negative regulator of abscisic acid signaling. *The Plant Cell*, 1999, 11, 1897–1909.

[85] Allen, G.J.; Kuchitsu, K.; Chu, S.P.; Murata, Y. & Schroeder, J.I. Arabidopsis abi1-1 and abi2-1 phosphatase mutations reduce abscisic acid-induced cytoplasmic calcium rises in guard cells. *The Plant Cell*, 1999, 11, 1785–1798.

[86] Murata, Y.; Pei, Z.M.; Mori, I.C. & Schroeder, J. Abscisic acid activation of plasma membrane $Ca^{2+}$ channels in guard cells requires cytosolic NAD(P)H and is differentially disrupted upstream and downstream of reactive oxygen species production in abi1-1 and abi2-1 protein phosphatase 2C mutants. *The Plant Cell*, 2001, 13, 2513–2523.

[87] Saez, A.; Apostolova, N.; Gonzalez-Guzman, M.; Gonzalez-Garcia, M.P.; Nicolas, C.; Lorenzo, O. & Rodriguez, P.L. Gain-of-function and loss-of-function phenotypes of the protein phosphatase 2C HAB1 reveal its role as a negative regulator of abscisic acid signalling. *The Plant Journal*, 2004, 37, 354–369.

[88] Kuhn, J.M.; Boisson-Dernier, A.; Dizon, M.B.; Maktabi, M.H. & Schroeder, J.I. The protein phosphatase AtPP2CA negatively regulates abscisic acid signal transduction in Arabidopsis, and effects of abh1 on AtPP2CA mRNA. *Plant Physiology*, 2006, 140, 127–139.

[89] Yoshida, T.; Nishimura, N.; Kitahata, N.; Kuromori, T.; Ito, T.; Asami, T.; Shinozaki, K. & Hirayama, T. ABA-hypersensitive germination3 encodes a protein phosphatase 2C (AtPP2CA) that strongly regulates abscisic acid signaling during germination among Arabidopsis protein phosphatase 2Cs. *Plant Physiology*, 2006, 140, 115–126.

[90] Nishimura, N.; Yoshida, T.; Kitahata, N.; Asami, T.; Shinozaki, K. & Hirayama, T. ABA-Hypersensitive Germination1 encodes a protein phosphatase 2C, an essential component of abscisic acid signaling in Arabidopsis seed. *The Plant Journal*, 2007, 50, 935–949.

[91] Koornneef, M.; L´eon-Kloosterziel, K.M.; Schwartz, S.H. & Zeevaart, J.A.D. The genetic and molecular dissection of abscisic acid biosynthesis and signal transduction in *Arabidopsis*. *Plant Physiology and Biochemistry*, 1998, 36, 83–89.

[92] Liotenberg, S.; North, H. & Marion-Poll, A. Molecular biology and regulation of abscisic acid biosynthesis in plants. *Plant Physiology and Biochemistry*, 1999, 37, 341–50.

[93] Xiong, L.; Ishitani, M.; Lee, H. & Zhu, J.K. The Arabidopsis *LOS5/ABA3* locus encodes a molybdenum cofactor sulfurase and modulates cold and osmotic stress responsive gene expression. *The Plant Cell*, 2001, 13, 2063–83.

[94] Cutler, S.R.; Rodriguez, P.L.; Finkelstein, R.R. & Abrams, S.R. Abscisic acid: Emergence of a core signaling network. *Annual Review of Plant Biology*, 2010, 61, 651–679.

[95] Kim, T.H.; Bo¨hmer, M.; Hu, H.; Nishimura, N. & Schroeder JI. Guard cell signal transduction network: Advances in understanding abscisic acid, CO2, and $Ca^{2+}$ signaling. *Annual Review of Plant Biology,* 2010, 61, 561–591.

[96] Hubbard, K.E.; Nishimura, N.; Hitomi, K.; Getzoff, E.D. & Schroeder, J.I. Early abscisic acid signal transduction mechanisms: newly discovered components and newly emerging questions. *Genes & Development*, 2010, 24, 1695–1708.

[97] Seki, M.; Umezawa, T.; Urano, K. & Shinozaki, K. Regulatory metabolic networks in drought stress responses. *Current Opinion in Plant Biology*, 2007, 10, 296–302.

[98] Zeevaart, J. A. D. & Creelman, R. A. Metabolism and physiology of abscisic acid. *Annual Review of Plant Physiology and Plant Molecular Biology*, 1988, 39, 439–473.

[99] Pierce, M. & Raschke, K. The relationship between abscisic acid accumulation and turgor. *Plant Physiology*, 1978, 61 (Suppl.), 25.

[100] Marin, E.; Nussaume, L.; Quesada, A.; Gonneau, M. & Sotta, B. Molecular identification of zeaxanthin epoxidase of *Nicotiana plumbaginifolia*, a gene involved in abscisic acid biosynthesis and corresponding to the ABA locus of *Arabidopsis thaliana*. *The EMBO Journal*, 1996, 15, 2331–1242.

[101] Tan, B.C.; Schwartz, S.H.; Zeevaart, J.A.D. & Mc-Carty, D.R. Genetic control of abscisic acid biosynthesis in maize. *Proceedings of the National Academy of Sciences of the United States of America*, 1997, 94, 12235–12340.

[102] Nambara, E. & Marion-Poll, A. Abscisic acid biosynthesis and catabolism. *Annual Review of Plant Biology*, 2005, 56, 165–185.

[103] Iuchi, S.; Kobayashi, M.; Taji, T.; Naramoto, M. & Seki, M. Regulation of drought tolerance by gene manipulation of 9-cis-epoxycarotenoid dioxygenase, a key enzyme in abscisic acid biosynthesis in Arabidopsis. *The Plant Journal*, 2001, 27, 325–33.

[104] Qin, X. & Zeevaart, J.A.D. The 9-*cis*epoxycarotenoid cleavage reaction is the key regulatory step of abscisic acid biosynthesis in water-stressed bean. *Proceedings of the National Academy of Sciences of the United States of America*, 1999, 96, 15354–15361.

[105] Thompson, A. J.; Jackson, A. C.; Symonds, R. C.; Mulholland, B. J.; Dadswell, A. R., Blake, P. S.; Burbidge, A. & Taylor, I. B. Ectopic expression of a tomato 9-*cis*-epoxycarotenoid dioxygenase gene causes over-production of abscisic acid. *The Plant Journal*, 2000, 23, 363–374.

[106] Liotenberg, S.; North, H. & Marion-Poll, A. Molecular biology and regulation of abscisic acid biosynthesis in plants. *Plant Physiology and Biochemistry*, 1999, 37, 341–350.

[107] Seo, M.; Peeters, A.J.M.; Koiwai, H.; Oritani, T. & Marion-Poll, A. The *Arabidopsis* aldehyde oxidase 3 (*AAO3*) gene product catalyzes the final step in abscisic acid biosynthesis in leaves. *Proceedings of the National Academy of Sciences of the United States of America*, 2000, 97, 12908–12913.

[108] Hirayama, T. & Shinozaki, K. Perception and transduction of abscisic acid signals: keys to the function of the versatile plant hormone ABA. *Trends in Plant Science*, 2007, 12, 343–351.

[109] Hirayama, T. & Shinozaki, K. Research on plant abiotic stress responses in the post-genome era: past, present and future. *The Plant Journal*, 2010, 61, 1041-1052.

[110] Yamaguchi-Shinozaki, K, & Shinozaki, K. Organization of cis-acting regulatory elements in osmotic- and cold-stress-responsive promoters. *Trends in Plant Science*, 2005, 10, 88-94.

[111] Shi, H.; Quintero, F.J.; Pardo J.M. & Zhu, J.K. The putative plasma membrane $Na^+/H^+$ antiporter SOS1 controls long distance $Na^+$ transport in plants. *The Plant Cell*, 2002, 14, 465–477.

[112] Apse, M.P.; Aharon, G.S.; Snedden, W.A. & Blumwald, E. Salt tolerance conferred by overexpression of a vacuolar NaCl/HCl-antiport in *Arabidopsis*. *Science*, 1999, 285, 1256–58.

[113] Zhu, J.K. Plant salt tolerance. *Trends in Plant Science*, 2001, 6, 66–71.

[114] Shinozaki, K. & Yamaguchi-Shinozaki, K. Gene expression and signal transduction in water-stress response. *Plant Physiology*, 1997, 115, 327–334.

[115] Uno, Y.; Furihata, T.; Abe, H.; Yoshida, R.; Shinozaki, K. & Yamaguchi-Shinozaki, K. Arabidopsis basic leucine zipper transcription factors involved in an abscisic acid-dependent signal transduction pathway under drought and high-salinity conditions. *Proceedings of the National Academy of Sciences of the United States of America*, 2000, 97, 11632–11637.

[116] Furihata, T.; Maruyama, K.; Fujita, Y.; Umezawa, T.; Yoshida, R. & Shinozaki, K. Abscisic acid-dependent multisite phosphorylation regulates the activity of a transcription activator AREB1. *Proceedings of the National Academy of Sciences of the United States of America*, 2006, 103, 1988-1993.

[117] Fujii, H. & Zhu, J. Arabidopsis mutant deficient in 3 abscisic acid-activated protein kinases reveals critical roles in growth, reproduction and stress. *Proceedings of the National Academy of Sciences of the United States of America*, 2009, 106, 8380-8385.

[118] Umezawa, T.; Yoshida, R; Maruyama, K. Yamaguchi-Shinozaki, K. & Shinozaki, K. SRK2C, a SNF1-related protein kinase 2, improves drought tolerance by controlling stress-responsive gene expression in Arabidopsis thaliana. *Proceedings of the National Academy of Sciences of the United States of America*, 2004, 101, 17306–17311.

[119] Kobayashi, Y.; Yamamoto, S.; Minami, H.; Kagaya, Y. & Hattori, T. Differential activation of the rice sucrose nonfermenting 1-related protein kinase2 family by hyperosmotic stress and abscisic acid. *The Plant Cell*, 2004, 16, 1163-1177.

[120] Kobayashi, Y.; Murata, M.; Minami, H.; Yamamoto, S.; Kagaya, Y. & Hobo, T. Abscisic acid activated SnRK2 protein kinases function in the gene-regulation pathway of ABA signal transduction by phosphorylating ABA response element-binding factors. *The Plant Journal*, 2005, 44, 939-949.

[121] Johnson, R.R.; Wagner, R.L.; Verhey, S.D. & Walker-Simmons, M.K. The abscisic acid-responsive kinase PKABA1 interacts with a seed-specific abscisic acid response element-binding factor, TaABF, and phosphorylates TaABF peptide sequences. *Plant Physiology*, 2002, 130, 837-846.

[122] Abe, H.; Urao, T.; Ito, T.; Sekic, M.; Shinozaki, K. & Yamaguchi-Shinozaki, K. Arabidopsis AtMYC2 (bHLH) and AtMYB2 (MYB) function as transcriptional activators in abscisic acid signaling. *The Plant Cell*, 2003, 15, 63–78.

[123] Jiang, C.; Iu, B. & Singh J. Requirement of a CCGAC cis-acting element for cold induction of the BN115 gene from winter *Brassica napus*. *Plant Molecular Biology*, 1996, 30, 679-684.

[124] Baker, S. S.; Wilhelm, K.S. & Thomashow, M.F. The 5'-region of Arabidopsis thaliana cor15a has cis-acting elements that confer cold-, drought- and ABA-regulated gene expression. *Plant Molecular Biology*, 1994, 24, 701-713.

[125] Liu, Q.; Kasuga, M.; Sakuma, Y.; Abe, H. & Miura, S. Two transcription factors, DREB1 and DREB2, with an EREBP/AP2 DNA-binding domain separate two cellular signal transduction pathways in drought- and low temperature– responsive gene expression, respectively, in Arabidopsis. *Plant Cell,* 1998, 10, 1391–406.

[126] Shinwari, Z.K.; Nakashima, K.; Miura, S.; Kasuga, M. & Seki, M. An Arabidopsis gene family encoding DRE/CRT binding proteins involved in low-temperature-responsive gene expression. *Biochemical and Biophysical Research Communications,* 1998, 250, 161–170.

[127] Gilmour, S.J.; Zarka, D.G.; Stockinger, E.J.; Salazar, M.P.; Houghton, J.M. & Thomashow, M.F. Low temperature regulation of the Arabidopsis CBF family of AP2 transcriptional activators as an early step in cold-inducedCORgene expression. *The Plant Journal,* 1998, 16, 433–442.

[128] Haake, V.; Cook, D.; Riechmann, J.L.; Pined,a O.; Thomashow, M.F. & Zhang, J.Z. Transcription factor CBF4 is a regulator of drought adaptation in Arabidopsis. *Plant Physiology*, 2002, 130, 639-648.

[129] Narusaka, Y.; Nakashima, K.; Shinwari, Z.K.; Sakuma, Y.; Furihata, T.; Abe, H.; Narusaka, M.; Shinozaki, K. & Yamaguchi-Shinozaki, K. Interaction between two cis-acting elements, ABRE and DRE, in ABA-dependent expression of Arabidopsis rd29A gene in response to dehydration and high-salinity stresses. *The Plant Journal,* 2003, 34, 137-148.

[130] Bhatnagar-Mathur, P.; Vadez, V. & Sharma, K.K. Transgenic approaches for abiotic stress tolerance in plants: retrospect and prospects. *Plant Cell Reports*, 2008, 27, 411-424.

[131] Century, K.; Reuber, T.L. & Ratcliffe, O.J. Regulating the regulators: the future prospects for transcription-factor-based agricultural biotechnology products. *Plant Physiology*, 2008, 147, 20-29.

[132] Nakashima, K.; Shinwari, Z.K.; Sakuma, Y.; Seki, M. & Miura, S. Organization and expression of two Arabidopsis DREB2 genes encoding DRE-binding proteins involved in dehydration- and high-salinity-responsive gene expression. *Plant Molecular Biology,* 2000, 42, 657–665.

[133] Stockinger, E.J.; Gilmour, S.J. & Thomashow, M.F. *Arabidopsis thaliana CBF1* encodes an AP2 domain-containing transcriptional activator that binds to the C-repeat/DRE, a cis-acting DNA regulatory element that stimulates transcription in response to low temperature and water deficit. .

*Proceedings of the National Academy of Sciences of the United States of America*, 1997, 94, 1035–40.

[134] Guiltinan, M.J.; Marcotte, W.R. & Quatrano, R.S. A plant leucine zipper protein that recognizes an abscisic acid response element. *Science,* 1990, 250, 267–271.

[135] Hobo, T.; Kowyama, Y. & Hattori, T. A bZIP factor, TRAB1, interacts with VP1 and mediates abscisic acid-induced transcription. *Proceedings of the National Academy of Sciences of the United States of America*, 1999, 96, 15348–15353.

[136] Nakagawa, H.; Ohmiya, K. & Hattori, T. A rice bZIP protein, designated OSBZ8, is rapidly induced by abscisic acid. *The Plant Journal,* 1996, 9, 217–227.

[137] Nantel, A. & Quatrano, R.S. Characterization of three rice basic/leucine zipper factors, including two inhibitors of EmBP-1 DNA binding activity. *The Journal of Biological. Chemistry,* 1996, 271, 31296–312305.

[138] Oeda, K.; Salinas, J. & Chua, N.H. A tobacco bZip transcription activator (TAF-1) binds to a G-box-like motif conserved in plant genes. *The EMBO Journal,* 1991, 10, 1793–17802.

[139] Finkelstein, R.R. & Lynch, T.J. The Arabidopsis abscisic acid response gene *ABI5* encodes a basic leucine zipper transcriptional factor. *The Plant Cell,* 2000, 12, 599–609.

[140] Bastola, D.R.; Pethe, V.V. & Winicov, I. Alfin1, a novel zinc-finger protein in alfalfa roots that binds to promoter elements in the salt-inducible MsPRP2 gene. *Plant Molecular Biology,* 1998, 38, 1123–1135.

[141] Yáñez, M.; Cáceres, S.; Orellana, S.; Bastías, A.; Verdugo, I.; Ruiz-Lara, S. & Casaretto, J.A. An abiotic stress-responsive bZIP transcription factor from wild and cultivated tomatoes regulates stress-related genes. *Plant Cell Reports*, 2009, 28, 1497-507.

[142] Hsieh, T.H.; Li, C.W.; Su, R.C.; Cheng, C.P.; Sanjaya Tsai, Y.C. & Chan, M.T. A tomato bZIP transcription factor, SlAREB, is involved in water deficit and salt stress response. *Planta*, 2010, 231, 1459-1473.

[143] Chaves, M.M.; Maroco, J.P. & Pereira, J.S. Understanding plant responses to drought - from genes to the whole plant. *Functional Plant Biology*, 2003, 30, 239-264.

[144] Tabaeizadeh, Z. Drought-induced responses in plant cells. *International Review of Cytology*, 1998, 182, 193-247.

[145] Rontein, D.; Basset, G. & Hanson, A.D. Metabolic engineering of osmoprotectant accumulation in plants. *Metabolic Engineering*, 2002, 4, 49-56.

[146] Crowe, J.H.; Carpenter, J.F. & Crowe, L.M. The role of vitrification in anhydrobiosis. *Annual Review of Physiology*, 1998, 60, 73-103.

[147] Hincha, D.K.; Zuther, E.; Hellwege, E.M. & Heyer, A.G. Specific effects of fructo- and gluco-oligosaccharides in the preservation of liposomes during drying. *Glycobiology*, 2002, 12, 103-110.

[148] Villadsen, D.; Rung, J.H. & Nielsen, T.H. Osmotic stress changes carbohydrate partitioning and fructose-2,6-bisphosphate metabolism in barley leaves. *Functional Plant Biology*, 2005, 32, 1033-1043.

[149] Tunnacliffe, A. & Wise, M.J. The continuing conundrum of the LEA proteins. *Naturwissenschaften*, 2007, 94, 791-812.

[150] Garay-Arroyo, A.; Colmenero-Flores, J.M.; Garciarrubio, A. & Covarrubias, A.A. Highly hydrophilic proteins in prokaryotes and eukaryotes are common during conditions of water deficit. *Journal of Biological Chemistry*, 2000, 275, 5668-5674.

[151] Goyal, K.; Walton, L.J. & Tunnacliffe, A. LEA proteins prevent protein aggregation due to water stress. *Biochemical Journal*, 2005, 388, 151-157.

In: Abiotic Stress: New Research
Editors: Nikhil Haryana and Shreya Punj
ISBN: 978-1- 61942-197-4


*Chapter 4*

# ABIOTIC STRESS RESPONSE IN PLANTS: INTEGRATIVE GENETIC PATHWAYS AND OVERLAPPING REACTIONS BETWEEN ABIOTIC AND BIOTIC STRESS RESPONSES

***Saúl Fraire-Velázquez*[1], *Lenin Sánchez-Calderón*[1] *and Salvador Guzmán-González*[2]**
[1]Unidad Académica de Ciencias Biológicas, Universidad Autónoma de Zacatecas, Guadalupe, Zacatecas, México
[2]Universidad de Colima, Facultad de Ciencias Biológicas y Agropecuarias, Km 40 Autopista Colima-Manzanillo, Tecomán, Colima, México

## ABSTRACT

In all ecosystems, plants continuously face environmental stress and consequently are forced to respond with defensive and adaptive strategies. These responses require the activation of several signaling pathways that induce expression of specific genes. An effective response requires that the biological system have the genetic background to support the necessary molecular players that permit the assembly of essential integrative genetic pathways. In the most complicated scenario, plants must contend against more than one abiotic stress, pest, or pathogen at the same time, forcing an integral and complete defense response to adjust plant physiology. Certain molecular players act as hubs or master regulators to integrate signals from different regulatory pathways activated by two or more types of abiotic stress or forms of biotic stress. Some defense and adaptive genes are transcriptionally regulated not only by a specific abiotic stress but also by biotic stress, exposing the existence of overlapping pathways. The genes shared between these networks appear to allow plants to prioritize their responses and ensure their survival by using their resources efficiently. The well-documented shared or convergence points in the response to abiotic and biotic stress in plants exhibit a superimposed complexity, as exemplified by induction of certain defense genes by biotic stress in presence only of a specific environmental condition (temperature and humidity). Interestingly, accumulating data strongly support the hypothesis that the convergence points between abiotic and biotic stress pathways also modulate the post-embryonic developmental program which is one of the most conspicuous adaptive strategies to cope with environmental stress. Pathogen attack or abiotic stresses such as nutrient scarcity alter cell division and cell differentiation processes, and consequently the plant architecture is modified. Plant growth regulators, such as auxins, cytokinins, ethylene, and jamonic acid, as well as reactive oxygen species, play crucial roles in the early steps of the convergence between these multiple stress signals. The advancement in powerful molecular tools, including transcriptome and proteome analysis, whole-genome sequencing, and bioinformatic studies are enabling disection of networks in abiotic and biotic signaling cascades and identification of the overlapping reactions and key factors that fulfill very important roles as integrative signals in plants.

## I. Introduction

Different forms of abiotic and biotic stress invariably limit the production of crops in the approximately 1.5 billion hectares of arable land in the world (UNESCO, 2009), a problem that is far from solved and that will be exacerbated constraining plant growth and productivity. Problems of lower yields due to drought are serious; irrigation will not be a long-term viable solution since water grows scarcer and large-scale desalinization is not yet viable. High and low temperatures, acid soils, and soils with high content of metal ions reduce productivity in many cropping areas. Solutions to these problems could be diverse, but a better understanding of the physiology and genetic of tolerance in plants to stress will be essential as we attempt to mitigate these limitations and to increase crops yields. Maximizing yields by breeding for resistance, or at least tolerance, to abiotic and biotic stress is of great importance. Increases in crop yields in developing countries are of high priority since these areas exhibit the highest levels of increasing population and consequently increasing demands on the food supply. Furthermore, the effects of climate change will be of major impact on lower latitudes that are occupied primarily by developing countries.

In the past two decades, great advances in plant molecular biology have permitted new discoveries about phenotype plasticity, defined as the capacity of a given plant genotype to produce different phenotypes in different environmental conditions (Sonia, 2010). Several genes that determine phenotypes of abiotic and biotic stress tolerance or resistance and involved signaling pathways have been identified. The study of gene function has been advanced firstly by whole genome sequences and bioinformatics approaches. At the same time, molecular biology techniques have been used to identify points of convergence where the signaling due to various stresses collides to activate specific response reactions. Formidable progress in the area of defense mechanisms to environmental stress in plants has been achieved, and the task now is to dissect the collections of players in the signaling pathways that transduce the messages from two or more cascades and identify the interconnections that exist among signaling routes. It is well known that when two or more types of stressing factors (biotic or abiotic; for instance, pathogen attack and heat stress or high relative humidity) coincide, the defense reaction to one stress impairs the reaction to the other. The knowledge of how biotic and abiotic environmental cues affect the final size and architecture of plants are progressing. We are beginning to understand in an integral way the biochemical, molecular, cellular, and physiological mechanisms involved in individual phenotypic plasticity. More effort is needed to draw the complete networks assembled in the intricate abiotic or biotic stress signaling responsible for control of the molecular, cellular, and physiological mechanisms of acclimation or adaptation.

## II. How Plants Integrate Regulatory Genetic Pathways

During 450 millions of years of their existence on the earth, the environment has imposed changing and stressing conditions on land plants. These conditions have forced the evolution of elaborate molecular systems that respond to dynamic situations with physiological changes that have allowed plants to adapt efficiently to life in the varied climates of the earth. These elaborate molecular systems have enabled the survival and adaptation in each plant species that reign over planet surface. In biological systems, it is well known that evolution has led to integrated and well engaged signaling pathways, that enable the organisms to respond to stressing conditions with the lesser waste of energy. An early event in plants after perception of environmental stresses is the activation of signaling cascades that in turn leads to the reprogramming of the profile of expressed genes including stress-responsive genes. In biotic stress, plant innate immunity is activated after perception of pathogen-associated molecular patterns (PAMPs) in charge of pattern recognition receptors (PRRs) or upon resistance (R) protein-mediated recognition of pathogen race-specific effector molecules (Figure 1). R proteins have been identified in many pathosystems, but only several PRRs have been described. In Arabidopsis several members of leucine-rich repeat with protein kinase activity have been found as PAMP receptors, which suggest members of this family function as pattern recognition receptors, and that heterologous expression of PAMP recognition systems is functional (Lacombe *et al.*, 2010).

A number of molecular players fulfill the relevant function of integrating signals from signaling pathways activated in responses to different types of biotic and abiotic stresses. A good example is Med25, a subunit of the *Arabidopsis thaliana* Mediator complex. Mediator is a phylogenetically conserved multiprotein complex of 20 to 30 subunits (depending of the organism) organized into three linked modules (the head, middle, and tail) and a detachable kinase module. The tail module is thought to interact primarily with DNA-bound transcription factors, while the head and middle modules bind to the C-terminal domain of RNA polymerase II (Kidd *et al.*, 2009). In this way Mediator promotes the assembly and activation of transcription complexes on core promoters, interacts with RNA Pol II in the initiation of transcription, and serves as a primary conduit of regulatory information from enhancers to promoters (Kuras *et al.*, 2003). Mediator is thought to be an integrator of regulatory signals that converge on promoters of stress-responsive genes, in this way, several subunits have been found functionally necessary in the transcription activated in response to diverse stress-specific signaling pathways (Kim *et al.*, 2006). Consistent with this, the Med25 subunit is a common target of three transcription factors involved in different pathways activated in response to salt, drought, and heat stress and also has a function in signaling that regulates flowering time in response to light conditions (Elfving *et al.*, 2011). That means that in these signaling cascades, Med25 (as part of Mediator) is the last signal receptor-transducer, just after these three transcription factors that have been found involved in stress response pathways, and just before the action of RNA Pol II (Figure 1). Med25 has a conserved activator-interacting domain (ACID) located in the 551-680 amino acid segment through which the molecule interacts with DNA-bound transcription regulatory factors. Transcription factors DREB2A, ZFHD1, and a MYB-like protein have been identified as Med25 interactors. DREB2A interacts with cis-acting dehydration-responsive element/C-repeat (DRE/CRT) involved in cold and drought stress–responsive gene expression and, in a constitutively active form, enhances drought tolerance (Sakuma *et al.*, 2006). The expression of ZFHD1 is induced by drought, high salinity and abscisic acid, and *Arabidopsis* plants overexpressing ZFHD1 have higher drought tolerance than wild-type plants (Tran *et al.*, 2007). MYB-like protein activates expression of transcripts specifically expressed in plants subjected to a combination of drought and heat stress (Rizhsky *et al.*, 2004). Furthermore, in *Arabidopsis*, the *PHYTOCHROME AND FLOWERING TIME1* (*PFT1*) gene, which encodes the MEDIATOR25 (Med25) subunit of Mediator, is required for jasmonate-dependent defense gene expression and resistance to leaf-infecting necrotrophic fungal pathogens (Kidd *et al.*, 2009) (Figure 1, biotic stress).

Another example of a molecular player that integrates signals is the Target of Rapamycin (TOR), a Ser/Thr kinase conserved in fungi, insects, mammals, and photosynthetic eukaryotes. Rapamycin is an anti-proliferative drug produced by *Streptomyces hygroscopicus* (Schmelzle and Hall, 2000) originally described as an antifungal agent (Vezina *et al.*, 1975). Rapamycin first binds to FK506-binding protein (FKBP12) and this complex inhibits the target TOR Ser/Thr kinase. In contrast to its activity in other eukaryotes, Rapamycin does not seem to affect TOR function in plants (Mahfouz *et al.*, 2006), although in the unicellular green alga *Chlamydomonas reinhardtii* TOR and FKBP12 homologs have been identified and characterized, and *Chlamydomonas* cells are sensitive to Rapamycin (Crespo *et al.*, 2005). TOR has a phosphatidylinositol 3-kinase domain in its C-terminal region, adjacent to the FKBP12-rapamycin-binding (FRB) domain through which it establishes interaction with the FKBP12-rapamycin complex. At the N-terminus, TOR has two blocks of HEAT motifs; these motifs are involved in protein-protein interactions (Andrade, 1995) Through the HEAT motifs, *Arabidopsis* TOR interacts with RAPTOR1 (a TOR regulatory protein), and RAPTOR1 regulates the activity of S6 kinase (S6K) in response to osmotic stress. S6K in turn phosphorylates ribosomal protein S6 (RPS6), through which regulates translation (Mahfouz *et al.*, 2006). TOR inactivation leads to a nutrient-starvation response, suggesting that TOR is involved in the reaction to nutrient deficiency (Barbet *et al.*, 1996). In *Arabidopsis*, AtFKBP12 interacts with AtFIP37, a protein involved in embryogenesis and endosperm development, placing AtFKBP12 in regulation of the cell cycle and in developmental processes (Vespa *et al.*, 2004). In yeasts and mammals it has been found that TOR is a central controller of cell growth (Schmelzle and Hall, 2000). In plants, TOR links embryonic development, cell growth, yield, stress resistance, and mRNA translation, integrating energy levels, nutrient availability, and stress information

(Abraham, 2005; Mahfouz *et al.*, 2006). In summary, TOR acts as a hub that optimizes cellular resources for growth, playing a central regulatory role in stress and growth signaling pathways. Med25 and TOR are two molecules that exemplify how plants have evolved to respond efficiently by integrating several pathways through key master regulators (Figure 1).

Signal integration from different stress signaling pathways is also mediated more simply than by Med25 and TOR through the protein-binding elements in the promoters of some defense genes. It is known for example that abiotic and biotic stresses such as salinity, drought, abscisic acid (ABA), and fungal inoculation induce similar patterns of expression of the members of the *14.3.3* gene family (*GF14b* and *GF14c*). The 14-3-3 proteins participate in cellular regulatory pathways as adapters, chaperones, activators, or repressors and execute important steps in signal transduction and metabolism (Chen *et al.*, 2006). These *GF14* genes share the characteristic of *cis*-elements in their promoter regions that are responsive to abiotic stress and pathogen attack (Figure 1). The *14-3-3s* family genes are also subject to regulation by certain transcript factors (Chen *et al.*, 2006). In rice, the promoter regions of *OsGF14b*, *c*, *e* and *f* genes contain low-temperature response elements (LTRE), whereas *OsGF14b*, *c*, *d* and *g* genes contain cooper response elements (CuRE), in addition to other elements related to abiotic stress such drought-responsive elements (DRE) and binding sites for MYB transcription factors widely distributed in promoter regions of *OsGF14s* (Yao *et al.*, 2007). Likewise, in *Arabidopsis* at least five of the 29 *cytochrome P450* genes are induced by abiotic and biotic stress including the pathogens *Alternaria brassicicola* or *Alternaria alternata*, paraquat, rose bengal, UV stress (UV-C), heavy metal stress ($CuSO_4$), mechanical wounding, drought, high salinity, low temperature, or hormones (salicylic acid, jasmonic acid, ethylene, and ABA). Five of these *cytochrome P450* genes (*CYP81D11*, *CYP710A1*, *CYP81D8*, *Cyp71B6* and *CYP76C2*) are co-induced by metal stress ($CuSO_4$), paraquat, salinity, ABA, and pathogen inoculation. In these *cytochrome P450* genes as in the *14.3.3* gene family, *cis*-acting elements include W-box (DNA binding sites for WRKY transcription factors), P-box (a positive *cis*-acting regulator of pathogen defense), and MYB recognition sites (Narusaka *et al.*, 2004). These data show clearly that plants have evolved diverse response elements to assemble special architectures in the regulatory regions of defense genes, which allow an integrated and efficient response against diverse environmental stressing constraints (Figure 1).

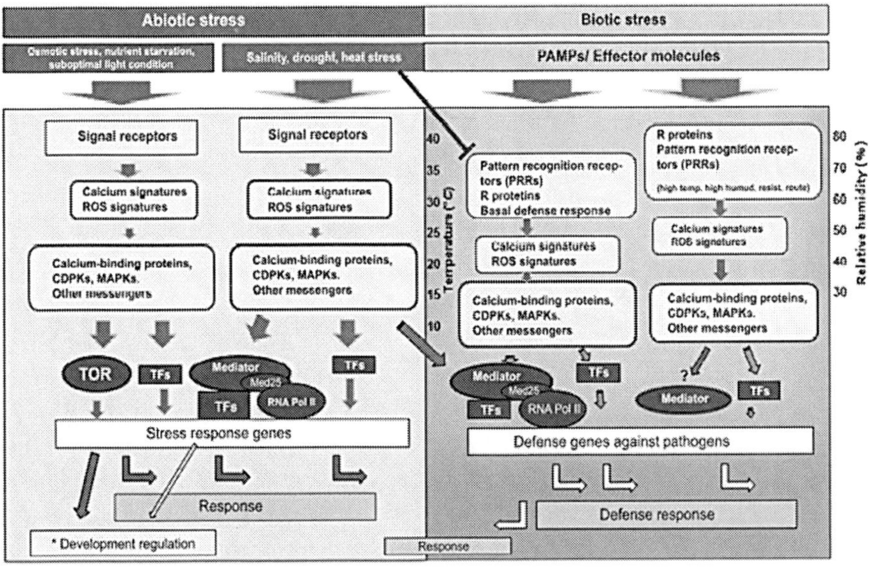


Figure 1. Signal cascade pathways in response to abiotic and biotic stress. In abiotic stress, besides classical transcription factors (TFs), TOR and Mediator (with Med25) as key regulator in the signal transduction to activate specific stress response genes. In biotic stress, cascade signaling through classical pattern recognition receptors (PRRs), R protein and Basal defense response, impaired by high temperature and high humidity, whereas several described R proteins act even in high temperature. Med25 also included in signal transduction to the expression of defense genes in response to pathogen. Furthermore, plant defense genes against pathogens, are induced in response to forms of abiotic stress. TOR and Med25 also involved in regulation of the development program.

## III. Superimposed Complexity in the Defense Mechanisms to Biotic Stress in Specific Abiotic Stress Conditions

The defense reaction to pathogen in some pathosystems exhibits another level of complexity. Variable environmental conditions are common during the plant life cycle and, as could be expected, stressing conditions, in the climate for example, exert another stress pressure in the host in a pathosystem. The pressure of this extra stress is manifested in the plant cells and at the end in the molecular mechanisms that plant implements to respond to these spliced stressing factors. In this direction, several studies with mutants of *Arabidopsis thaliana* show a resistance phenotype that depends on environmental conditions, specifically temperature and relative humidity; under low humidity and cold temperatures the resistance is enhanced, and in high relative humidity and high temperature the resistance phenotypes are suppressed (Yoshioka *et al.*, 2001; Mosher *et al.*, 2010). Similarly, in *A. thaliana* and *Nicotiana benthamiana*/*Pseudomonas syringae* pathosystems, the basal resistance and (*R*) gene-mediated resistance are reduced at moderately elevated temperatures (Wang *et al.*, 2009). Temperature sensitivity is thus present in defense mediated by different classes of R proteins, and also, often correlates with lesser hypersensitive response (Figure 1, biotic stress side), an observable fact not attributable to pathogen-secreted effectors. The negative effect of high temperature on the defense reaction is a common phenomenon reported in plant disease resistance against biotrophic and hemibiotrophic pathogens (Wang *et al.*, 2009). Indeed, studies on plant-bacterial pathosystems indicate that the expression of virulence genes tends to increase below temperatures of optimal microorganism growth, and several bacterial effectors are secreted preferentially in a range of temperature between 18 and 22°C (Van Dijk *et al.*, 1999).

That means that the advent of increasing environmental temperatures may impair (*R)* gene-mediated disease resistance and result in more extensive disease development, although there are examples of *R* genes that exhibit biological performance in the opposite direction with higher efficacy at high temperatures (Uauy *et al.*, 2005; Fu *et al.*, 2009; Webb *et al.*, 2010) (Figure 1, biotic stress side). The detrimental temperature effect on *R*-gene-mediated and basal disease resistance indicates that genetic pathways activated in response to biotic stress are modulated by environmental factors to variable levels. The biological basis of high temperature could be explained by considering the impact of temperature on protein denaturation and aggregation and on nucleic acid denaturation; this in turn affects protein-protein or protein-nucleic acid interactions along all steps of the cascade pathway, for example in the avirulence product-R protein recognition and in the transcription factor-promoter sequence interaction. Thus, higher temperatures result in lower affinity between interacting factors, less specific recognition of pathogen molecules, and less optimal interactions between interacting players downstream in the signaling cascade, including the transcription factor and the response element in the respective gene, and finally non-optimal *R* gene expression.

In an environment with periods of high temperature, plants besides to cope with pathogen challenge, must respond with a physiology change to adapt and survive to the stressing condition. Plants could respond with a process of acclimation termed thermotolerance; in this process the changes at molecular level imply new profiles of expressed genes with emphasis on stress-related proteins. A well-characterized response to heat stress is the expression of heat shock proteins

(HSPs) that work as molecular chaperones assisting in folding, ATP-dependent refolding, intracellular distribution, assembly and degradation of proteins, and stabilizing partially unfolded states for thermoprotection (Lee and Vierling, 2000; Wegele *et al.*, 2004). A characteristic effect of high temperature is damage to cellular structures and membrane dysfunction. Heat stress affects cell physiology by increasing the fluidity of membrane lipids, a structure whose interior lipid bilayer is normally highly fluid. In fact, normal cell function requires membrane lipid bilayers that are largely fluid, a common state at physiological temperatures. In accordance with this, cytoplasmic membrane function depends critically on the physical state of lipid bilayer, making it susceptible to changes in environmental temperature (Mansilla *et al.*, 2004). The cell membrane is a structure for anchorage of membrane proteins such as signal receptors, ion-channels, and the machinery of translocation. Furthermore, high temperature results in high transpiration and this in turn may increase the number of particles dissolved in the cell water content (i.e., an increase in solute potential, $\psi_s$) and decrease the water potential ($\psi_w$). Peripheral proteins located on the membrane surface commonly are water-soluble with mostly hydrophilic surfaces. Conditions that disrupt ionic and hydrogen bond interactions would affect the biological function of R proteins located in the cytoplasmic membrane or in the cytosol, restraining the activation of the defense response at the first step in the signal cascade.

## IV. Modulating Program of Post-Embryonic Development by Biotic and Abiotic Cues

All land-dwelling plants are attached to the soil and they cannot avoid the unfavorable conditions prevailing in its surroundings, as a consequence, they always are exposed to multivariate environmental cues. Plants evolved to perceive and integrate these biotic and/or abiotic cues and adjust their growth and development according to multivariate inputs. This capacity of plant genotype for producing different phenotypes under different environmental conditions, called phenotypic plasticity, allows plants to change dramatically their final body appearance to cope with environmental heterogeneity (Valladares *et al.*, 2007; Capron *et al.*, 2009; Sonia, 2010). This capacity is maintained even in mature plants in part due to their typical apical growth.

As part of their sessile lifestyle, plants show a particular kind of growth and development. Their mature cells are enclosed in a polysaccharide rigid matrix which maintains them attached to their neighbors. Unlike metazoans, plant development is given by the progressive addition of new cells through the activity of local populations of totipotent cells termed meristems (Meyerowitz, 2002; Nakajima and Benfey, 2002; Taiz and Zeiger, 2010). The plant development is guided by a genetically controlled program and it can be divided in two main stages: a) embryonic development (ED) and b) post-embryonic development (PED). During the ED, through a suite of highly regulated and reproducible stages, the fertilized egg cell rises into an embryo. In the embryo, the primary meristems, body axes and major tissue layers are established (Jurgens, 2001; Nakajima and Benfey, 2002; Willemsen and Scheres, 2004; Capron *et al*., 2009). Almost all the body of the mature plant is generated during the PED, which begins during germination, as soon as the mitotic activity of meristems commences. Both the primary shoot meristem (SM) and primary root meristem (RM) occupy opposite ends of the main body axis. The first, located in the upper part, gives rise to new organs such as leaves, stems, secondary meristems, flowers and fruits. The second, at the bottom part, originates the root system (Willemsen and Scheres, 2004; Vernoux *et al*., 2010; Teotia and Lamb, 2011). The genetic program of PED regulates: i) primary meristems activity, ii) *de novo* formation of secondary meristems and organs, and iii) cell elongation, determining the final plant size and shape (Sánchez-Calderón *et al*., 2005; Taiz and Zeiger, 2010). Due to the high plasticity of the program of PED, the size and shape of plant can change according to environmental cues.

In soil, biotic and abiotic components are distributed heterogeneously, and there, supplies of nutrients and water are very limited, localized, and variable. Also, a broad range of chemical and physical processes occurs due to intrinsic soil characteristics and the action of biotic factors (Lynch, 1995; McCully, 1995). As soon as the primary root emerges from the seed, it has to grow through these hostile media. As growth goes on, *de novo* lateral roots are formed from the primary root. Primary and lateral roots form the root system (RS), which has an spatial configuration called root architecture (Lynch, 1995). According to RS morphology and architecture, the root-soil interaction area, called rhizosphere, can increase exponentially. Is in this area where the intricate and multivariate interactions among plants, chemical and physical soil components and macro and microorganisms take part. This complexity of biotic and abiotic cues must to be sensed and integrated by plants in order to adjust their program of PED. For example, as nutrients are distributed in a patching heterogeneous pattern, plants change their PED during their life in order to find the nutrient-rich regions, and in fact they develop more total root biomass in richer regions. Moreover, plants selectively destine more resources to those roots developing in those regions with increasing nutrient level availability, despite having other roots growing in already richer regions (Hodge, 2004; Shemesh *et al.*, 2010). Nutrients such as nitrogen (N), phosphorus (P), potassium (K) and sulfur (S) have been reported to modulate the program of PED (Zhang and Forde, 1998; Forde and Lorenzo, 2001; Williamson *et al.*, 2001; Kutz *et al.*, 2002; Lopez-Bucio *et al.*, 2002; Ashley *et al.*, 2006).

When growing under limiting P conditions, plants of *Arabidopsis thaliana* show dramatic changes in root architecture such as reduction in primary root length, increased formation of lateral roots and greater formation of root hairs. P deficiency induces a change of program of PED from determinate to indeterminate. Typically, on indeterminate PED, the newly formed root cells are added by the mitotic activity of primary meristem. These cells then get away from the meristem and increase their length, and the elongation process ends when the cells start to differentiate. When plants are P starved, cell division in the primary root meristems gradually reduces and the cells start to prematurely differentiate until total inhibition of cell elongation and loss of meristematic activity occur (meristem exhaustion). At the end, root tips change their physiological characteristics and the exhausted meristem becomes a structure which takes part in P uptake (Williamson *et al.*, 2001; Lopez-Bucio *et al.*, 2002; Sánchez-Calderón *et al.*, 2005). In this process, root tips locally detect P deficiency, this response being mediated by at least LPR multicopper oxidase genes (Sánchez-Calderón *et al.*, 2005; Svistoonoff *et al.*, 2007). Recently, iron (Fe) has been reported to play a role as well in the control of these PED reprogramming (Ward *et al.*, 2008). This change of root architecture is due to the fact that, in both meristematic and elongation areas, the content of reactive oxygen species (ROS) is reduced as long as the determined PED goes on. The *low phosphorus insensitive 4* (*lpi4*) mutant does not show the typical P deficiency growth phenotype, neither the ROS reduction in root tips. Interestingly, the addition of jasmonate (JA) to low P availability medium rescues the wild type (WT) growth phenotype, suggesting that this phytohormone, commonly related to biotic stress, is taking part over developmental root system changes in response to nutritional stress (Chacón-López *et al.*, 2011), and could be a link integrating both abiotic and biotic stress response signals. However, addition of JA in media also inhibits the primary root growth in WT plans independently of the media P content (Berger *et al.*, 1996). Crosstalk can be inferred by analyzing if a specific mutant shares components or check-points regarding responses to abiotic and biotic stresses. For example, the *coi1* (coronatine insensitive1), *jar1* (jasmonate resistant1) and *jin1* (jasmonate insensitive1) mutants are affected in JA signaling (Feys *et al.*, 1994; Berger *et al.*, 1996; Staswick *et al.*, 2002). *COI1*encodes a F-box protein, an integral component of multi-protein complexes implicated in ubiquitination. JAR1, a JA amino acid synthetase, is required to activate JA for signaling. *JIN1*encodes a transcription factor of the helix-loop-helix type which transcriptionally regulates the expression of JA-responsive genes (Xie *et al.*, 1998; Lorenzo *et al.*, 2004; Staswick and Tiryaki, 2004). Genetic damage in these *loci* modifies the sensitivity of these mutants to bacterial and fungal pathogens and insect attack (Feys *et al.*, 1994; Anderson *et al.*, 2004; Lorenzo *et al.*, 2004; Dombrecht *et al.*, 2007); besides, it reduces the root development sensitivity to JA, clearly indicating that *COI1, JAR1* and *JIN1* are necessary for root development and defense responses. Future work using *coi1, jar1* and *jin1* is necessary to confirm the crosstalk among changes in root development specifically induced by P starvation and plant defense responses. Finally, in the shoot, the shade avoidance syndrome and some specific defense responses are regulated by integrated jasmonate and light signals (Kazan and Manners, 2011).

## Conclusion

Until recently, it was thought that plant responses to biotic or abiotic stresses involved a unidirectional signaling. However, in the past decade our knowledge about the sophisticated signaling pathways that have evolved to deal with environmental changes has increased dramatically. We now understand how plants sense and integrate the multivariate biotic and abiotic stressors allowing intricate crosstalk among different signaling pathways in the response to the many environmental stresses. In this context, the proteins Med25 and TOR are key molecular players in this integration of multiple stressor signals. The identification of these control points shared in signaling networks, and work toward an understanding of how they are regulated and how they in turn regulate downstream responses is of paramount importance, as we seek to determine how plants modulate their development, physiology and metabolism in a wide spectrum of phenotypic responses. We expect that in the near future, this research will enable the development of new generations of tolerant crop varieties. These new strains should have better performance under stressful environmental constraints, leading to higher yields and productivity, necessary as we seek to respond to the food demands of the increasing world population.

## ACKNOWLEDGMENTS

We thank M. Ibarra-Cortes for excellent English edition work. Writing of this paper has been made possible by a financial support from Secretaría de Educación Pública-Programa de Mejoramiento al Profesorado (SEP-PROMEP), México. Grants UAZ-PTC-104 and Red de Cuerpos Académicos: Biotecnología para el desarrollo de una Agricultura sustentable, UAZ-CA 138.


## REFERENCES


Abraham, R. T. (2005) TOR signaling: an odyssey from cellular stress to the cell growth machinery. *Curr. Biol.* 15: R139-141

Anderson, J. P., Badruzsaufari, E., Schenk, P. M., Manners, J. M., Desmond, O. J., Ehlert, C., Maclean, D. J., Ebert, P. R., Kazan, K. (2004) Antagonistic Interaction between Abscisic Acid and Jasmonate-Ethylene Signaling Pathways Modulates Defense Gene Expression and Disease Resistance in Arabidopsis. *The Plant Cell Online* 16: 3460-3479

Andrade, M. A., Bork, P. (1995) HEAT repeats in the Huntington's pathway. disease protein. *Nat. Genet.* 11, 115–116.

Ashley, M. K., Grant, M., Grabov, A. (2006) Plant responses to potassium deficiencies: a role for potassium transport proteins. *Journal of Experimental Botany* 57: 425-436

Barbet, N. C., Schneider, U., Helliwell, S. B., Stansfield, I., Tuite, M. F., Hall, M. N. (1996) TOR controls translation initiation and early G1 progression in yeast. *Mol. Biol.* Cell 7: 25-42

Berger, S., Bell, E., Mullet, J. E. (1996) Two Methyl Jasmonate-Insensitive Mutants Show Altered Expression of AtVsp in Response to Methyl Jasmonate and Wounding. *Plant Physiology* 111: 525-531

Capron, A., Chatfield, S., Provart, N., Berleth, T. (2009) Embryogenesis: Pattern Formation from a Single Cell. *The Arabidopsis Book*: 1-28

Crespo, J. L., Diaz-Troya, S., Florencio, F. J. (2005) Inhibition of target of rapamycin signaling by rapamycin in the unicellular green alga Chlamydomonas reinhardtii. *Plant Physiol.* 139: 1736-1749

Chacón-López, A., Ibarra-Laclette, E., Sánchez-Calderón, L., Gutiérrez-Alanis, D., Herrera-Estrella, L. (2011) Global expression pattern comparison between <em>low phosphorus insensitive 4</em> and WT Arabidopsis reveals an important role of reactive oxygen species and jasmonic acid in the root tip response to phosphate starvation. *Plant Signaling and Behavior* 6: 382-392

Chen, F., Li, Q., Liangxian, S., He, Z. (2006) The Rice 14-3-3 Gene Family and its Involvement Responses to Biotic and Abiotic Stress. *DNA Research:* 53-63

Chen, F., Li, Q., Sun, L., He, Z. (2006) The rice 14-3-3 gene family and its involvement in responses to biotic and abiotic stress. *DNA Res*. 13: 53-63

Dombrecht, B., Xue, G. P., Sprague, S. J., Kirkegaard, J. A., Ross, J. J., Reid, J. B., Fitt, G. P., Sewelam, N., Schenk, P. M., Manners, J. M., Kazan, K. (2007) MYC2 Differentially Modulates Diverse Jasmonate-Dependent Functions in Arabidopsis. *The Plant Cell Online* 19: 2225-2245

Elfving, N., Davoine, C., Benlloch, R., Blomberg, J., Brannstrom K., Muller, D., Nilsson, A., Ulfstedt, M., Ronne, H., Wingsle, G., Nilsson, O., Bjorklund, S. (2011) The Arabidopsis thaliana Med25 mediator subunit integrates environmental cues to control plant development. *Proc. Natl. Acad. Sci.* U S A 108: 8245-8250

Feys, B., Benedetti, E., Penfold, C. N., Turner, J. G. (1994) Arabidopsis Mutants Selected for Resistance to the Phytotoxin Coronatine Are Male Sterile, Insensitive to Methyl Jasmonate, and Resistant to a Bacterial Pathogen. *The Plant Cell Online* 6: 751-759

Forde, B., Lorenzo, H. (2001) The nutritional control of root development. *Plant and Soil* 232: 51-68

Fu, D., Uauy, C., Distelfeld, A., Blechl, A., Epstein, L., Chen, X., Sela, H., Fahima, T., Dubcovsky, J. (2009) A kinase-START gene confers temperature-dependent resistance to wheat stripe rust. *Science* 323: 1357-1360

Hodge, A. (2004) The plastic plant: root responses to heterogeneous supplies of nutrients. *New Phytologist* 162: 9-24

Jurgens, G. (2001) Apical-basal pattern formation in Arabidopsis embryogenesis. *EMBO J.* 20: 3609-3616

Kazan, K., Manners, J. M. (2011) The interplay between light and jasmonate signalling during defence and development. *Journal of Experimental Botany*

Kidd, B., Edgar, C., Kumar, K., Aitken E., Schenk, P., Manners, J., Kazan, K. (2009) The Mediator Complex Subunit PFT1 Is a Key Regulator of Jasmonate-Dependent Defense in Arabidopsis. *Plant Cell* 21: 2237-2252

Kim, S., Xu, X., Hecht, A., Boyer, T. G. (2006) Mediator is a transducer of Wnt/beta-catenin signaling. *J. Biol. Chem*. 281: 14066-14075

Kuras, L., Borggrefe, T., Kornberg, R. D. (2003) Association of the Mediator complex with enhancers of active genes. *Proc. Natl. Acad. Sci*. U S A 100: 13887-13891

Kutz, A., Müller, A., Hennig, P., Kaiser, W. M., Piotrowski, M., Weiler, E. W. (2002) A role for nitrilase 3 in the regulation of root morphology in sulphur-starving Arabidopsis thaliana. *The Plant Journal* 30: 95-106

Lacombe, S., Rougon-Cardoso, A., Sherwood, E., Peeters, N., Dahlbeck, D., Van Esse, H. P., Smoker, M., Rallapalli, G., Thomma, B. P., Staskawicz, B., Jones, J. D., Zipfel, C. (2010) Interfamily transfer of a plant pattern-recognition receptor confers broad-spectrum bacterial resistance. *Nat. Biotechnol.* 28:365-9

Lee, G. J., Vierling, E. (2000) A small heat shock protein cooperates with heat shock protein 70 systems to reactivate a heat-denatured protein. *Plant Physiol.* 122: 189-198

Lopez-Bucio, J., Hernandez-Abreu, E., Sanchez-Calderon, L., Nieto-Jacobo, M. F., Simpson, J., Herrera-Estrella, L. (2002) Phosphate Availability Alters Architecture and Causes Changes in Hormone Sensitivity in the Arabidopsis Root System. *Plant Physiol.* 129: 244-256

Lorenzo, O., Chico, J. M., Sánchez-Serrano, J. J., Solano, R. (2004) JASMONATE-INSENSITIVE1 Encodes a MYC Transcription Factor Essential to Discriminate between Different Jasmonate-Regulated Defense Responses in Arabidopsis. *The Plant Cell Online* 16: 1938-1950

Lynch, J. (1995) Root Architecture and Plant Productivity. *Plant Physiology* 109: 7-13

Mahfouz, M. M., Kim, S., Delauney, A. J., Verma, D. P. (2006) Arabidopsis TARGET OF RAPAMYCIN interacts with RAPTOR, which regulates the activity of S6 kinase in response to osmotic stress signals. *Plant Cell* 18: 477-490

Mansilla, M. C., Cybulski, L. E., Albanesi, D., De Mendoza, D. (2004) Control of membrane lipid fluidity by molecular thermosensors. *J. Bacteriol.* 186: 6681-6688

McCully, M. (1995) How Do Real Roots Work? (Some New Views of Root Structure). *Plant Physiology* 109: 1-6

Meyerowitz, E. M. (2002) Plants Compared to Animals: The Broadest Comparative Study of Development. *Science* 295: 1482-1485

Mosher, S., Moeder, W., Nishimura, N., Jikumaru, Y., Se-Hwan, J., Urquhart, W., Klessig, D., Kim, S., Nambara, E., Yoshioka, K. (2010) The Lesion-Mimic Mutant cpr22 Shows Alterations in Abscisic Acid Signaling and Abscisic Acid Insensitivity in a Salicylic Acid-Dependent Manner. *Plant Physiol.* 152: 1901-1913

Nakajima, K., Benfey, P. N. (2002) Signaling In and Out: Control of Cell Division and Differentiation in the Shoot and Root. *The Plant Cell Online* 14: S265-S276

Narusaka, Y., Narusaka, M., Seki, M., Umezawa, T., Ishida, J., Nakajima, M., Enju, A., Shinozaki, K. (2004) Crosstalk in the responses to abiotic and biotic stresses in Arabidopsis: Analysis of gene expression in cytochrome P450 gene superfamily by cDNA microarray *Plant Mol. Biol.* 55: 327-342

Rizhsky, L., Liang, H., Shuman, J. Shulaev, V., Davletova, S., Mittler, R. (2004) When defense pathways collide. The response of Arabidopsis to a combination of drought and heat stress. *Plant Physiol.* 134: 1683-1696

Sakuma, Y., Maruyama, K., Qin, F., Osakabe, Y., Shinozaki, K., Yamaguchi-Shinozaki, K. (2006) Dual function of an Arabidopsis transcription factor DREB2A in water-stress-responsive and heat-stress-responsive gene expression. *Proc. Natl. Acad. Sci.* U S A 103: 18822-18827

Sánchez-Calderón, L., López-Bucio, J., Chacón-López, A., Cruz-Ramírez, A., Nieto-Jacobo, F., Dubrovsky, J. G., Herrera-Estrella, L. (2005) Phosphate Starvation Induces a Determinate Developmental Program in the Roots of Arabidopsis thaliana. *Plant and Cell Physiology* 46: 174-184

Schmelzle, T., Hall, M. N. (2000) TOR, a central controller of cell growth. *Cell* 103: 253-262

Shemesh, H., Arbiv, A., Gersani, M., Ovadia, O., Novoplansky, A. (2010) The Effects of Nutrient Dynamics on Root Patch Choice. PLoS ONE 5: e10824

Sonia, E. S., (2010) Plant developmental responses to the environment: eco-devo insights. *Current Opinion in Plant Biology* 13: 96-101

Staswick, P. E., Tiryaki, I. (2004) The Oxylipin Signal Jasmonic Acid Is Activated by an Enz1yme That Conjugates It to Isoleucine in Arabidopsis. *The Plant Cell Online* 16: 2117-2127

Staswick, P. E., Tiryaki, I., Rowe, M. L. (2002) Jasmonate Response Locus JAR1 and Several Related Arabidopsis Genes Encode Enzymes of the Firefly Luciferase Superfamily That Show Activity on Jasmonic, Salicylic, and Indole-3-Acetic Acids in an Assay for Adenylation. *The Plant Cell Online* 14: 1405-1415

Svistoonoff, S., Creff, A., Reymond, M., Sigoillot-Claude, C., Ricaud, L., Blanchet, A., Nussaume, L., Desnos, T. (2007) Root tip contact with low-phosphate media reprograms plant root architecture. *Nat. Genet.* 39: 792-796

Taiz, L., Zeiger, E. (2010) *Plant Physiology*, Ed Fifth. Sinauer Associates, U.S.A

Teotia, S., Lamb, R. S. (2011) RCD1 and SRO1 are necessary to maintain meristematic fate in Arabidopsis thaliana. *Journal of Experimental Botany* 62: 1271-1284

Tran, L. S., Nakashima, K., Sakuma, Y., Osakabe, Y., Qin, F., Simpson, S. D., Maruyama, K., Fujita, Y., Shinozaki, K., Yamaguchi-Shinozaki, K. (2007) Co-expression of the stress-inducible zinc finger homeodomain ZFHD1 and NAC transcription factors enhances expression of the ERD1 gene in Arabidopsis. *Plant J.* 49: 46-63

Uauy, C., Brevis, J. C., Chen, X., Khan, I., Jackson, L., Chicaiza, O., Distelfeld, A., Fahima, T., Dubcovsky, J. (2005) High-temperature adult-plant (HTAP) stripe rust resistance gene Yr36 from Triticum turgidum ssp. Dicoccoides is closely linked to the grain protein content locus Gpc-B1. *Theoret. App. Genetics* 112: 97–105

UNESCO (2009) 3rd UN World Water *Development Report*, 2009. In http://www.unesco.org/water/wwap/

Valladares, F., Gianoli, E., Gómez, J. M. (2007) Ecological limits to plant phenotypic plasticity. *New Phytologist* 176: 749-763

Van Dijk, K., Fouts, D. E., Rehm, A. H., Hill, A. R., Collmer, A., Alfano, J. R. (1999) The Avr (effector) proteins HrmA (HopPsyA) and AvrPto are secreted in culture from Pseudomonas syringae pathovars via the Hrp (type III) protein secretion system in a temperature- and pH-sensitive manner. *J. Bacteriol.* 181: 4790-4797

Vernoux, T., Besnard, F., Traas, J. (2010) Auxin at the Shoot Apical Meristem. *Cold Spring Harbor Perspectives in Biology* 2

Vespa, L., Vachon, G., Berger, F., Perazza, D., Faure, J. D., Herzog, M. (2004) The immunophilin-interacting protein AtFIP37 from Arabidopsis is essential for plant development and is involved in trichome endoreduplication. *Plant Physiol.* 1238-1292

Vezina, C., Kudelski, A., Sehgal, S. N. (1975) Rapamycin (AY-22,989), a new antifungal antibiotic. I. Taxonomy of the producing streptomycete and isolation of the active principle. *J. Antibiot.* (Tokyo) 28: 721-726

Wang, Y., Bao, Z., Zhu, Y., Hua, J. (2009) Analysis of Temperature Modulation of Plant Defense Against Biotrophic Microbes. *MPMI 22*: 498–506

Wang, Y., Bao, Z., Zhu, Y., Hua, J. (2009) Analysis of temperature modulation of plant defense against biotrophic microbes. *Mol. Plant-Microbe Interact.* 22: 498-506

Ward, J. T., Lahner, B., Yakubova, E., Salt, D. E., Raghothama, K. G. (2008) The Effect of Iron on the Primary Root Elongation of Arabidopsis during Phosphate Deficiency. *Plant Physiology* 147: 1181-1191

Webb, K. M., Ona, I., Bai, J., Garrett, K. A., Mew, T., Vera Cruz, C. M., Leach, J. E. (2010) A benefit of high temperature: increased effectiveness of a rice bacterial blight disease resistance gene. *New Phytol.* 185: 568-576

Wegele, H., Muller, L., Buchner, J. (2004) Hsp70 and Hsp90--a relay team for protein folding. *Rev. Physiol. Biochem. Pharmacol.* 151: 1-44

Willemsen, V., Scheres, B. (2004) Mechanisms of Pattern Formation in Plant Embryogenesis. *Annual Review of Genetics* 38: 587-614

Williamson, L. C., Ribrioux, S. P. C. P., Fitter, A. H., Leyser, H. M. O. (2001) Phosphate Availability Regulates Root System Architecture in Arabidopsis. *Plant Physiol.* 126: 875-882

Xie, D., Feys, B., James, S., Nieto-Rostro, M., Turner, J. (1998) COI1: An Arabidopsis Gene Required for Jasmonate-Regulated Defense and Fertility. *Science* 280: 1091-1094

Yao, Y., Du, Y., Jiang, L., Liu,.J. Y. (2007) Molecular Analysis and Expression Patterns of the 14-3-3 Gene Family from Oryza Sativa. *J. Bioch. Mol. Biol.* 40: 349-357

Yoshioka, K., Kachroo, P., Tsui, F., Sharma, S., Shah, J., Klessig, D. (2001) Environmentally-sensitive, SA-dependent defense response in the cpr22 mutant of Arabidopsis. *Plant Journal* 26: 447-459

Zhang, H., Forde, B. G. (1998) An Arabidopsis MADS Box Gene That Controls Nutrient-Induced Changes in Root Architecture. *Science* 279: 407-409

In: Abiotic Stress: New Research
Editors: Nikhil Haryana and Shreya Punj
ISBN: 978-1- 61942-197-4


*Chapter 5*

# CIRCADIAN REGULATION OF ABIOTIC STRESS TOLERANCE IN *CAENORHABDITIS ELEGANS*

***Andres Romanowski, Maria Eugenia Goya, Maria Laura Migliori, Sergio H. Simonetta and Diego A. Golombek***

Laboratorio de Cronobiología, Departamento de Ciencia y Tecnología, Universidad Nacional de Quilmes, Buenos Aires, Argentina;
National Research Council, Argentina

## ABSTRACT

Circadian rhythms are ubiquitously found in nature. These rhythms are generated and sustained by a conserved time-keeping system that coordinates the biochemistry, physiology and behavior of an organism with daily changes in the environment. There is increasing evidence that this coordination contributes to the overall health of organisms. Hence, it is of great importance to understand how the circadian clock is able to transduce environmental signals to internal processes. One of the outputs of the circadian system in invertebrates, vertebrates, bacteria and plants is the stress resistance behavior. In this work we focus on the circadian variation in tolerance to several abiotic stressors in the nematode *Caenorhabditis elegans*. Circadian and daily variations in response to osmotic-, oxidative-

and cyanide-mediated stress have been described at the behavioral and molecular level.

Overall, the biological clock helps *C. elegans* cope with naturally occurring rhythms in abiotic stressors present in its natural habitat. It provides tools for preparing the organism to defend itself by predicting when the defense mechanisms are most needed.

## INTRODUCTION

We live in a world that spins around its axis with a period of approximately 24 hours. Because of this we are subjected to daily rhythms in several environmental variables, such as the succession of night and day, temperature, UV-light incidence and humidity. This fact has shaped the evolution of the organisms that live on this world and most, if not all, have developed a very accurate time-keeping system to cope with and predict such daily variations. This system is known as the biological clock and it helps an organism know "when" it is situated in a particular day. In this way it predicts the periodic fluctuations in the environment and orchestrates changes at the physiological, behavioral and molecular level that allow the organism to be prepared to withstand those changes.

When the period of a rhythm is close to 24 hours it is called a circadian rhythm (from the latin words *circa* = around and *diem* = day) (Figure 1). Rhythms with frequencies faster than a day are called *ultradian* rhythms (e.g.: respiratory rhythm, cardiac rhythm) and those with frequencies running slower than a day are called *infradian* rhythms (e.g.: hibernation rhythm) [1].

It has been demonstrated that circadian rhythms depend on environmental cues and can be entrained by different stimuli such as light or temperature. When an organism is kept under constant environmental conditions most physiological variables remain rhythmic, which suggests the presence of an endogenous clock. In this way, the biological clock can be described as a three component system: an exogenous component or *input* (also known as *Zeitgeber*), a biological clock or *core*, and the biological rhythms or *output* (Figure 2). However, the diverse components interact with one another and, for example, the *Zeitgeber* may directly affect a biological variable bypassing the clock (this is known as *masking*) and; also the biological rhythms can themselves feedback into the biological clock mechanism.

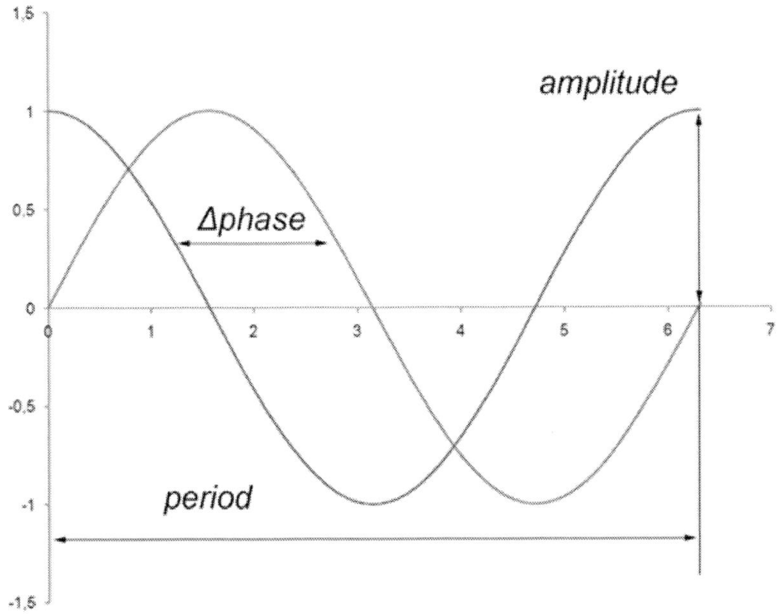


Figure 1. Properties of a biological rhythm. The equation that describes this waveform is $y(t) = A. \cos(wt + \theta)$, where $A$ is the amplitude; $w$ is the angular velocity and is related to the period of the wave; $t$ is time; and $\theta$ is the phase angle. In this figure both waves share the same amplitude and period, but have a phase angle difference of 90° or $\pi/2$.

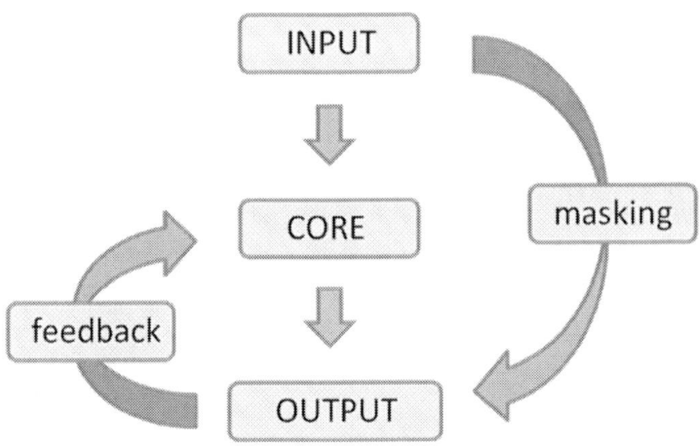


Figure 2. Basic diagram of the circadian system. The main components of the clock are the input or *zeitgeber*; the core, which is the biological clock itself; and the output. When an input influences the output directly, bypassing the core, this is called masking. The outputs can themselves give feedback to the core.

Even though light and temperature are the main *zeitgebers*, other environmental variables may affect the clock, including pH changes; redox changes; osmolarity; etc., some of which are abiotic stressors. In this sense, different environmental cues might guide or entrain rhythmic cellular responses to environmental stress. Since many stress response machineries are conserved among many different species it is possible to find similar patterns of circadian regulation across different organisms. This is the case, for example, for the heat shock proteins (mainly HSP70 and HSP16), which are clock-controlled in *Synechocystis*, *Neurospora*, *Gonyaulax* and *Drosophila* [2, 3], at transcriptional or induction levels. Rhythms in antioxidant enzymes such as superoxide dismutase, catalase, ascorbate peroxidase, haloperoxidase and glutathione peroxidase are regulated at transcriptional or post-translational levels in many different organisms spanning various kingdoms [4]. Among the diverse experimental models extensively used in chronobiological research, we propose to add one of the most powerful model systems characterized mainly in developmental biological and genetics.

Indeed, *Caenorhabditis elegans* is a soil-dwelling nematode widely used as a model organism in several areas of research, such as aging, molecular biology, genomics and neurobiology. To name some of its advantages, it does not take up much housing space (it is 1mm long as an adult), has a short generation time (around 4 days at 20 °C), is hermaphroditic, lays around 300 eggs in its lifetime, feeds on *E. coli*, its full genome and that of several related species has been sequenced, there are lots of mutant strains available, it is easy to perform RNAi screens and its neuronal wiring diagram has been characterized [5, 6]. In addition, very recently *C. elegans* has been used to study chronobiology and some circadian behaviors have already been described [7-12].

In this chapter we will discuss the basis of circadian rhythms, current knowledge about circadian rhythms in *C. elegans*, different types of abiotic stress, and the rhythms in abiotic stress tolerance that have been described in *C. elegans*.

## Circadian Rhythms

To consider a rhythm as truly circadian it must obey the following rules: a) it must persist in constant conditions with a period close to 24 hours; b) it can be entrained by exogenous stimuli, called synchronizing agents and; c) it must be temperature compensated [1, 13].

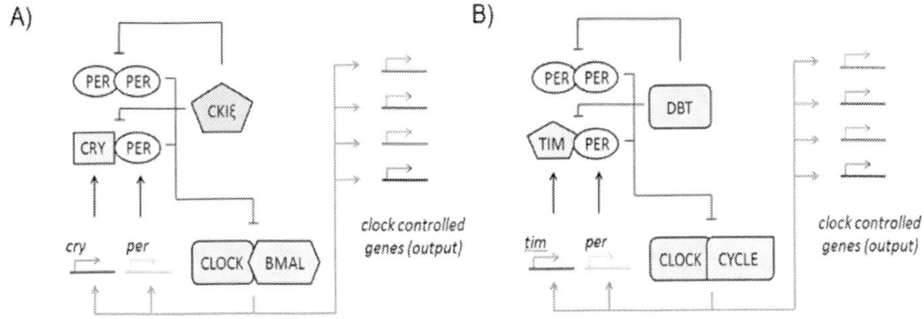


Figure 3. Basic diagram of the molecular mechanism of the circadian clock. A) Mammalian clock; B) *Drosophila melanogaster's* clock. PER = period; CRY = cryptochrome; CKIξ = Casein Kinase 1 ξ; TIM = timeless; DBT = doubletime.

The key gene and protein components of the clock have been established in many organisms, such as mammals [14], the fruit fly *Drosophila melanogaster* [15], fungi [16], plants [17] and bacteria [18]. The molecular clock consists basically of positive and negative elements that regulate each other in a negative feedback loop. Nonetheless, even though the mechanism of the clock of several organisms is the same, the components are not necessarily conserved, suggesting that the circadian clock must have appeared several independent times during evolution, and also speaks of the importance of having a way to predict cyclic events. The basic outline of the mammalian and insect clocks can be seen in Figure 3.

In mammals, the main components of the clock are CLOCK, BMAL, PER and CRY. During the day, a CLOCK:BMAL heterodimer activates the transcription of the *per* and *cry* genes. Once translated, PER:PER and PER:CRY dimers form and translocate to the nucleus by nightfall. These dimers interfere with CLOCK:BMAL and repress the transcription of the *per* and *cry* genes. In turn when the PER-CRY dimer is degraded, CLOCK:BMAL is able to induce the transcription of PER and CRY thus closing the loop. In addition, several post-transcriptional events fine-tune the circadian molecular machinery.

For example, CKIε also regulates the clock by fulfilling three roles: 1) tagging PER monomers for their degradation; 2) helping translocate PER:PER and PER:CRY dimers to the nucleus; and 3) it is involved in the degradation of these dimers once they have fulfilled their repressive role [14].

In the case of *Drosophila* the process is very similar, but with slight differences. The dimer that activates the transcription of *per* is CLOCK:CYCLE. Instead of activating CRY this dimer activates TIMELESS (TIM) and the TIM:PER dimers are then accumulated in the cytoplasm to then enter the nucleus by nightfall and interfere with the action of CLOCK:CYCLE, and hence the transcription of *per* and *tim*.

The fly homologue of CKIε, DOUBLETIME, is in charge of tagging the PER monomers in the cytoplasm. When morning arrives, CRY is activated and induces the degradation of TIMELESS in the PER:TIM dimers, whereas DBT induces the degradation of PER in the PER:TIM dimers [15].

These genes by themselves do not determine the actual period of the clock; it is the existence of neural networks coupled as a system that will ultimately determine the period of the biological clock.

## THE BIOLOGICAL CLOCK OF *C. ELEGANS*

The clock mechanism of *C. elegans* has yet to be elucidated. Bioinformatics studies have revealed that *C. elegans* shares many homologues to insect and mammalian clock genes [19, 20] (a list of *C. elegans* homologues to Drosophila clock genes can be seen in Table 1) but the core clock homologues appear to have heterochronic functions in the nematode [21-23]. Nevertheless, several circadian and daily rhythms have been described in this nematode. The first report of circadian rhythmicity in *C. elegans* was shared by two independent reports that appeared in 2002. One of the reports, by Saigusa et al., described the presence of circadian rhythms in the swimming behavior (locomotor activity) of adult nematodes [11]. The other one, by Kippert et al., described rhythms in hyperosmotic stress tolerance of L1 larvae [10]. Since then, several rhythms have been described: rhythms in hyperosmotic and oxidative stress tolerance [9]; rhythms in tolerance to biotic stress mediated by *Pseudomonas fluorescens* and HCN exposure [24]; rhythms in metabolic variables such as oxygen consumption, feeding, pharyngeal pumping, changes in defecation rate [7]; rhythms in melatonin synthesis and aaNAT activity [25]. Also, the melatonin receptors [26] and homologues to *Drosophila* PDF peptides and receptors [27-30] have been described.

Most of the papers about the circadian biology of *C. elegans* describe circadian outputs. However all of the results relied in discrete sampling of nematodes and none of them offered a continuous readout of a circadian variable. This would be important to allow the screening of clock mutants. In an effort to do so, our lab has designed an automated tracking system to measure the locomotor activity of nematodes and other small animals [31, 32]. With this system we were able to determine that the endogenous period of the *C. elegans'* clock is about 23.9 hours and that *C. elegans* entrains to light:dark(LD) and warm:cold temperature (Tt) cycles [8]. Interestingly, a *lin-42* mutant, homologous to the gene *period*, showed an endogenous period of about 25 hours and when rescued returned to a wild-type period [8].

**Table 1. *C. elegans*' homologues to *D. melanogaster*'s clock genes**

| Gene name | Protein name | *C. elegans* gene | Homology (%) | E -value | Loop |
|---|---|---|---|---|---|
| **per** | Period | lin-42 | 58% | 2.00E-14 | 1st loop |
| **tim** | Timeless | tim-1 | 52% | 5.00E-03 | 1st loop |
| **clk** | Clock | aha-1 | 40% | 9.00E-10 | 1st loop |
| **cyc** | Cycle | aha-1 | 53% | 6.00E-52 | 1st loop |
| **vri** | Vrille | atf-2 | 85% | 2.00E-09 | 2nd loop |
| **pdp1e** | PAR domain protein 1e (isoform J) | ces-2 | 50% | 5.00E-06 | 2nd loop |
| | PAR domain protein 1e (isoform D) | ces-2 | 50% | 5.00E-06 | |
| **cwo** | Clockwork orange | aha-1 | 50% | 6.00E-04 | 3rd loop |
| **dbt** | Doubletime | kin-20 | 85% | 1.00E-141 | modulation of per/tim |
| **ck2** | Casein Kinase 2 (alpha) | kin-3 | 83% | 1.00E-147 | modulation of per/tim |
| **ck2** | Casein Kinase 2 (beta) | kin-10 | 83% | 5.00E-90 | |
| **sgg** | Shaggy | gsk-3 | 84% | 1.00E-145 | modulation of per/tim |
| **pp1** | PP1-9C (flw) | gsp-1 | 90% | 1.00E-160 | modulation of per/tim |
| | PP1-13C | gsp-2 | 94% | 1.00E-169 | |
| | PP1-87B | gsp-2 | 96% | 1.00E-171 | |
| | PP1-96A | gsp-2 | 92% | 1.00E-192 | |
| **pp2a** | Protein Phosphatase 2a | pptr-2 | 84% | 0.00E+00 | modulation of per/tim |
| **slmb** | Supernumerary limbs | lin-23 | 77% | 0.00E+00 | modulation of per/tim |
| **cry1** | Cryptochrome 1 | no hits found | na | na | input/tim degradation |
| **rh1 (ninaE)** | Rhodopsin 1 | dop-4 | 47% | 5.00E-21 | input |
| **rh6** | Rhodopsin 6 | npr-2 | 43% | 3.00E-14 | input |
| **rh5** | Rhodopsin 5 | f41e4.7 | 47% | 6.00E-13 | input |
| **to** | Takeout | c13a2.6 | 38% | 6.60E-01 | output |
| **pdf** | Pigment Dispersing Factor | pdf-1 | 18aa/180 | na | output |
| | | nlp-37 | 18aa/89 | na | |
| **nmo** | Nemo | lit-1 | 79% | 1.00E-140 | CLK & PER modulation |

**Table 1. (Continued)**

| Gene name | Protein name | *C. elegans* gene | Homology (%) | E -value | Loop |
|---|---|---|---|---|---|
| **slo** | Slowpoke | slo-1 | 72% | 0.00E+00 | output |
| **na** | Narrow abdomen | C27F2 | 64% | 0 | output |
| **Ir** | Inward rectifier (Ir-PB, PC, PD & PE) | M02A10.2 | 68% | 1.00E-101 | output |
| **jet** | Jetlag (jet-PA & PB) | C02F5.7A | 44% | 3.00E-13 | involved in light-dep TIM degradation |
| **csn4** | COP9 signalosome (CSNA-PA & PB) | Y55F3AM.15 | 57% | 3.00E-59 | involved in light-dep TIM degradation |
| **creb2** | cAMP Response Element (crebB-17A) | crh-1 | 68% | 1.00E-12 | core clock |
| **pka** | Protein Kinase A (Pka-C1-PA, PB & PC) | ZK909.2h | 88% | 1.00E-169 | output |
| **ebony** | Beta-alanyl conjugase synthetase (CG3331-PA) | C41A3.1 | 41% | 2.00E-15 | output |
| **lrk** | Lark (PA. PB, PC, PD & PE) | ZK863.7 | 66% | 3.00E-19 | output |
| **nocte** | Nocte-PA & PB | no hits found | na | na | temperature entrainment |
| **norpA** | Phospholipase C (norpA-PA PB, PC, PD & PE) | B0348.4A | 65% | 1.00E-124 | temperature entrainment |
| **fer2** | Transcription factor (CG5952-PB) | F48D6.3 | 73% | 3.00E-27 | identity sLNvs |

More recently, a genome-wide analysis of circadian transcripts entrained to LD and Tt cycles has been reported and several genes have been found to be expressed in a circadian manner. The datasets described that 1817 transcripts were Tt driven; 775 were LD driven; 286 were entrained to Tt cycles; and 406 were entrained to LD cycles. The comparison of the datasets showed that there were 107 transcripts in common in the driven datasets and only 2 in the entrained dataset (*wdr-5.3*, a homologue of the WDR5 member of the histone methyltransferase complex known to associate with mammalian PER1; and *Y102A5C.6*, a pseudogene). The latter result is very important because if *C. elegans* had a unique clock that could be entrained by both light and temperature the clock components should appear in this analysis. This could mean one of two things. One is that *C. elegans* could have two distinct clocks and the other one is that masking could have occurred because of non-cycling tissues, which means that clock mRNAs could have been missed. However, another important result is that the mRNA of clock genes homologues does not seem to cycle in any of the conditions tested. This means that the clock mechanism of *C. elegans* could be an entirely new one. The authors also propose a role for TAX-2, a CNG channel involved in light and temperature transduction. They showed that a mutant strain of *tax-2* (strain PR671) had severely affected transcriptional rhythms. They tested this by means of a temperature entrainable GFP reporter (*nlp36*p::*gfp*) and of three randomly chosen light entrainable transcripts. This indicates that tax-2 is needed to transduce these *Zeitgebers* to the clock. It is interesting to say that *lite-1* mutants which code for the LITE-1 photoreceptor, that is required for light responses in the nematode, did not appear to affect light entrainment [12].

## CIRCADIAN RHYTHMS AND ABIOTIC STRESS IN *C. ELEGANS*

The concept of stress is sometimes used in an imprecise manner, which can lead to confusion. We use the term stress to name any external factor that affects the nematodes in a negative way. In particular, we will be discussing environmental (or abiotic) factors, although biotic factors (such as interactions with pathogenic bacteria) can also exert stress. In most cases (not only in *C. elegans*) stress is measured in relation to the organism survival.

The ability of an organism to cope with stressful stimuli is called *stress tolerance* and is also referred to as *stress resistance*. Both terms are used in the scientific literature, although *stress tolerance* is preferred.

*C. elegans*' natural habitat is soil, where they are exposed to many different types of stressors, such as osmotic stress, heat stress, cold stress, UV light radiation, oxidative stress, etc. This fact, together with the characteristics already mentioned, makes this species an excellent organism to study interactions with environmental stressors. This statement is reflected in the wide variety of literature available on stress tolerance to heavy metals [33-40], osmotic stress [41-44], desiccation [45, 46], oxidative stress [47-49], oxygen deficiency [50-52], temperature [53, 54] and UV-light [55-58].

Kippert et al. were the first to describe a rhythmic tolerance to environmental stress. They studied hyperosmotic stress tolerance to 1.3M NaCl in populations of L1 larvae and found a rhythm that could be entrained to LD cycles and continued cycling in constant darkness. They found that nematodes had maximum tolerance or survival during the day and that the minimum tolerance was in the middle of the night. They also estimated the endogenous period to be about 22 hours and that the rhythms were temperature compensated with a $Q_{10}$ of about 1.06 [10]. A few years later and in line with this research, our lab described rhythms in stress tolerance to osmotic (350mM NaCl) and oxidative stress (50mM $H_2O_2$) in adult populations of nematodes. We confirmed previous results and also showed that osmotic stress tolerance peaked during the day and had a minimum at night in adults as well, in both LD and constant darkness conditions. Oxidative stress tolerance, on the other hand, peaked at night and had a minimum during the day. We estimated the endogenous periods to be about 20.6 and 20.3 hours in osmotic and oxidative rhythms, respectively (Figure 4). We also identified two stress-related genes that varied in expression between night and day, *gpdh-1* and *gpx* (*C11E4.1*) had higher levels of expression during the night [9]. These results confirmed the fact that *C. elegans* shows rhythms in stress tolerance to at least oxidative and osmotic stressors and hinted that stress-related genes could also be regulated in a daily manner.

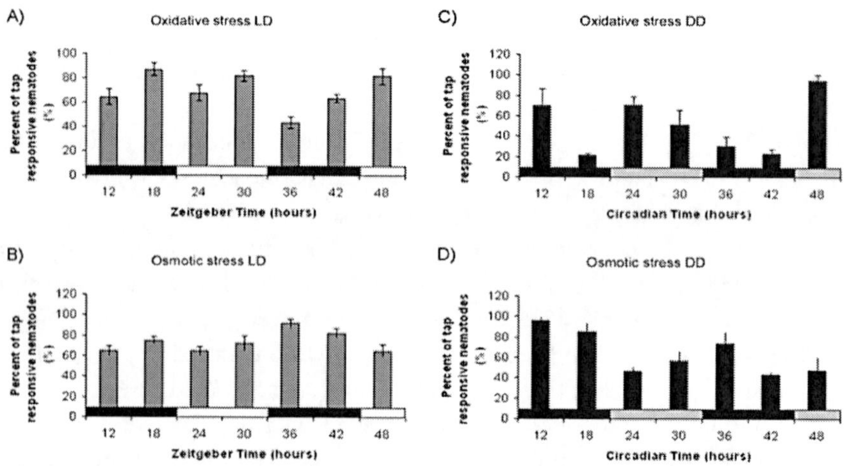


Figure 4. Rhythmic stress tolerance in *C. elegans*. The figures show the stress tolerance of population of nematodes at different time points. A) Oxidative stress tolerance in light:dark (LD) conditions; B) Osmotic stress tolerance in LD; C) Oxidative stress tolerance in constant darkness (DD conditions); D) Osmotic stress tolerance in DD conditions.

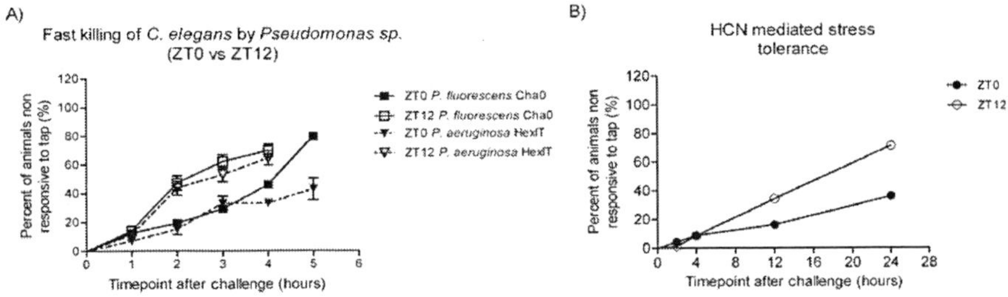


Figure 5. Biotic stress is sometimes sensed as abiotic. A) Daily variation in tolerance to *Pseudomonas aeruginosa* HexIT (inverted triangle) and *Pseudomonas fluorescens* (square). Nematode populations are more rapidly killed at lights off, ZT12 (open figure), than at lights on, ZT0 (filled figure). The secondary metabolite associated with this type of killing by these bacteria is hydrogen cyanide. B) Daily variation in tolerance to HCN. As expected, nematodes are more rapidly killed by HCN when exposed at ZT12 (open circle) than at ZT0 (filled circle).

The recent work by van der Linden et al. has shed new light in the circadian biology of *C. elegans* and we might be tempted to compare their datasets to available microarray datasets of nematodes exposed to abiotic stressors to determine if the stress response pathways vary in a circadian manner. When we compared the LD and Tt entrained transcripts [12] to the list of genes induced after 1 hour of hyperosmotic stress [41], we found that several genes involved in the osmotic stress response pathways oscillated with a period of 24 h (Table 2). The comparison of such datasets will help us further understand how the stress response machinery is regulated in a circadian manner.

It is interesting that in *C. elegans* the response to pathogens sometimes seems to be due to a response to the stress exerted by the infection and not through recognition of the pathogen itself [59]. For example, the response to *Bacillus thuringiensis* occurs through the interaction with the Cry5B or pore forming toxin (PFT) that the pathogen releases and involves the p38 and JNK MAPK pathways, partially overlapping the response to stress by heavy metals [60]. This is also the case when analyzing the response to the interaction with *Pseudomonas aeruginosa* PAO1, which paralyze and kill nematodes through the production of hydrogen cyanide (HCN). The response to this stressors was found to be mediated through the HIF pathway [61, 62]. In this sense, we aimed to study if the response to biotic stress also showed daily variations and found that the biotic stress mediated by either *Pseudomonas aeruginosa* HexIT or *Pseudomonas fluorescens* CHA0, which is also soil-dwelling as *C. elegans*, did indeed show a daily pattern of variation. We found that nematodes were more susceptible to a type of killing mediated by toxins, termed "fast paralytic killing", during the night

## Table 2. Genes upregulated during hyperosmotic stress that are also circadianly regulated

| LDsynchronized (LD and DD) | WCsynchronized (WC and CC) | LD driven | WC driven | | | only DD (entrained in LD) | only CC (entrained in WC) |
|---|---|---|---|---|---|---|---|
| R151.5 | *none* | Y40B10A.2 | Y51A2D.13 | F28A12.4 | Y119D3B.13 | Y102A5C.6 | F15E6.8 |
| R193.2 | | T27D12.1 | R06B10.3 | ZC443.3 | F44E5.5 | C49F5.1 | Y45G12C.2 |
| C12C8.2 | | R06B10.3 | K08F8.4 | T09F5.9 | F42D1.2 | F21F8.4 | M02D8.4 |
| M03B6.2 | | C49G7.5 | T10G3.3 | ZC416.6 | K02B7.2 | ZK484.2 | K02G10.7 |
| Y40D12A.2 | | T24C4.4 | F53B2.8 | ZK1127.10 | Y4C6B.6 | T01B11.7 | F47B10.2 |
| R02E12.6 | | T27E4.9 | C49G7.5 | K07E8.3 | C17C3.1 | K07F5.9 | C23H3.7 |
| C33C12.3 | | C24G6.6 | ZK484.2 | F31F7.1 | M60.2 | ZC416.6 | Y102A5C.6 |
| D2096.3 | | ZK418.7 | C24G6.6 | K08H10.1 | F08F3.7 | | F11A6.2 |
| | | R09B5.8 | F15E11.15 | M03B6.2 | K09H11.7 | | C30G12.2 |
| | | C24B9.9 | C50D2.6 | C17H12.8 | R193.2 | | B0213.4 |
| | | W02D3.7 | C24B9.3 | R02E12.6 | T13B5.1 | | Y51H4A.7 |
| | | C52A10.1 | F15E11.12 | Y38E10A.15 | W07B8.1 | | Y41C4A.11 |
| | | Y41C4A.11 | C48B4.1 | K11H12.4 | K11G9.6 | | B0213.5 |
| | | C14C11.3 | Y43F8B.2 | C09H5.2 | F21C10.9 | | C35D10.14 |
| | | R11F4.1 | F56C3.9 | Y71H10A.2 | F28H7.3 | | |
| | | T01B11.7 | Y19D10B.7 | T27E4.9 | C47A4.2 | | |
| | | F10E9.12 | C15C8.3 | F28F8.2 | F21C10.10 | | |
| | | C28C12.5 | F11A3.1 | F49C12.7 | F13A7.11 | | |
| | | F08F3.7 | K11H3.1 | Y46D2A.2 | F22B8.6 | | |
| | | Y94H6A.10 | F46G10.6 | F09C8.1 | C07H4.1 | | |
| | | F15E6.3 | Y40D12A.2 | C25F6.3 | T22G5.7 | | |
| | | F08F8.5 | C40H1.7 | R09B5.9 | C47A10.1 | | |
| | | T09F5.9 | F53F1.4 | F33D4.1 | Y58A7A.3 | | |
| | | C07H4.1 | Y94H6A.10 | ZK550.2 | F23F12.12 | | |
| | | Y58A7A.5 | F21F8.4 | W02D3.7 | F07C3.9 | | |
| | | F44G3.2 | F36H1.5 | F44E5.4 | C04F5.7 | | |
| | | C17H12.8 | T07E3.4 | ZK550.6 | C33C12.3 | | |
| | | | W04E12.8 | T15B7.4 | C31C9.1 | | |

(Figure 5A). By using different types of bacterial mutants we determined that the major toxin involved in the nematodes mortality was hydrogen cyanide, which we could catalogue as an abiotic stressor. When we subjected populations of nematodes to HCN stress we found that nematodes were more susceptible to HCN during the night than during the day (Figure 5B), which was in accordance with the results we obtained when using the biotic stress assay [24].

# CONCLUSION

Being *C. elegans* an established model organism for various areas of research, and because of the several environmental stressors it encounters in its natural habitat, it is a great research tool to study organism-environment interactions.

Several circadian rhythms have been described and the field of *C. elegans* chronobiology seems to be flourishing. Rhythms in response to oxidative, osmotic and cyanide exposure have been described and point out that *C. elegans* has an internal clock that somehow regulates its physiology, metabolism and behavior to be prepared to endure and predict daily changes in environment variables. However, the molecular mechanism of the circadian clock of *C. elegans* still remains elusive.

The tools available to us in the genomic era will allow us to further investigate which stress-related genes and pathways are under circadian control and help us elucidate how the circadian clock of *C. elegans* regulates its interaction with the environment. However, we also have to bear in mind that circadian regulation does not only occur at the transcriptional level. Several forms of posttranscriptional regulation have been described, i.e. regulation of splicing, capping and polyadenylation, surveillance, and stability [63, 64]; and also posttranslational regulation, i.e. phosphorylation, sumoylation, acetylation and degradation of protein products [64, 65]. Indeed, environmental stimuli, including potential stressors, might also affect the elusive circadian clock of *C. elegans* by targeting such post-translational mechanisms, which add up to the complexity of the circadian system.

# REFERENCES

[1] Golombek D. *Cronobiologia Humana:* Editorial Universitaria de Quilmes; 2002.

[2] Rensing L, Monnerjahn C. Heat shock proteins and circadian rhythms. *Chronobiology international.* 1996;13(4):239-50. Epub 1996/10/01.

[3] Ceriani MF, Hogenesch JB, Yanovsky M, Panda S, Straume M, Kay SA. Genome-wide expression analysis in Drosophila reveals genes controlling circadian behavior. *The Journal of neuroscience: the official journal of the Society for Neuroscience.* 2002;22(21):9305-19. Epub 2002/11/06.

[4] Hardeland R, Coto-Montes A, Poeggeler B. Circadian rhythms, oxidative stress, and antioxidative defense mechanisms. *Chronobiology international.* 2003;20(6):921-62. Epub 2003/12/19.

[5] Riddle DL. *C. elegans II. Plainview,* N.Y.: Cold Spring Harbor Laboratory Press; 1997. xvii, 1222 p. p.

[6] Wood WB. *The Nematode Caenorhabditis elegans. Cold Spring Harbor,* N.Y.: Cold Spring Harbor Laboratory; 1988. xiii, 667 p. p.

[7] Migliori ML, Simonetta SH, Romanowski A, Golombek DA. Circadian rhythms in metabolic variables in Caenorhabditis elegans. *Physiology & behavior.* 2011;103(3-4):315-20. Epub 2011/02/15.

[8] Simonetta SH, Migliori ML, Romanowski A, Golombek DA. Timing of locomotor activity circadian rhythms in Caenorhabditis elegans. *PloS one.* 2009;4(10):e7571. Epub 2009/10/28.

[9] Simonetta SH, Romanowski A, Minniti AN, Inestrosa NC, Golombek DA. Circadian stress tolerance in adult Caenorhabditis elegans. *Journal of comparative physiology A, Neuroethology, sensory, neural, and behavioral physiology.* 2008;194(9):821-8. Epub 2008/07/24.

[10] Kippert F, Saunders DS, Blaxter ML. *Caenorhabditis elegans has a circadian clock. Current biology:* CB. 2002;12(2):R47-9. Epub 2002/01/31.

[11] Saigusa T, Ishizaki S, Watabiki S, Ishii N, Tanakadate A, Tamai Y, et al. Circadian behavioural rhythm in Caenorhabditis elegans. *Current biology:* CB. 2002;12(2):R46-7. Epub 2002/01/31.

[12] van der Linden AM, Beverly M, Kadener S, Rodriguez J, Wasserman S, Rosbash M, et al. *Genome-wide analysis of light- and temperature-entrained circadian transcripts in Caenorhabditis elegans.* PLoS biology. 2010;8(10):e1000503. Epub 2010/10/23.

[13] Dunlap J, Loros J, Decoursey P. *Chronobiology: Biological Timekeeping:* Sinauer; 2003.

[14] Leloup JC, Goldbeter A. Toward a detailed computational model for the mammalian circadian clock. *Proceedings of the National Academy of Sciences of the United States of America.* 2003;100(12):7051-6. Epub 2003/05/31.

[15] Claridge-Chang A, Wijnen H, Naef F, Boothroyd C, Rajewsky N, Young MW. Circadian regulation of gene expression systems in the Drosophila head. *Neuron.* 2001;32(4):657-71. Epub 2001/11/24.

[16] Dunlap JC, Loros JJ. The neurospora circadian system. *Journal of biological rhythms.* 2004;19(5):414-24. Epub 2004/11/10.

[17] Millar AJ, Kay SA. The genetics of phototransduction and circadian rhythms in Arabidopsis. *BioEssays: news and reviews in molecular, cellular and developmental biology*. 1997;19(3):209-14. Epub 1997/03/01.

[18] Kondo T, Ishiura M. *The circadian clock of cyanobacteria. BioEssays: news and reviews in molecular, cellular and developmental biology.* 2000;22(1):10-5. Epub 2000/01/29.

[19] Temmerman L, Meelkop E, Janssen T, Bogaerts A, Lindemans M, Husson SJ, et al. C. elegans homologs of insect clock proteins: a tale of many stories. *Annals of the New York Academy of Sciences.* 2011;1220:137-48. Epub 2011/03/11.

[20] Hasegawa K, Saigusa T, Tamai Y. Caenorhabditis elegans opens up new insights into circadian clock mechanisms. *Chronobiology international.* 2005;22(1):1-19. Epub 2005/05/04.

[21] Banerjee D, Kwok A, Lin SY, Slack FJ. Developmental timing in C. elegans is regulated by kin-20 and tim-1, homologs of core circadian clock genes. *Developmental cell.* 2005;8(2):287-95. Epub 2005/02/05.

[22] Jeon M, Gardner HF, Miller EA, Deshler J, Rougvie AE. Similarity of the C. elegans developmental timing protein LIN-42 to circadian rhythm proteins. *Science.* 1999;286(5442):1141-6. Epub 1999/11/05.

[23] Ambros V. Control of developmental timing in Caenorhabditis elegans. *Current opinion in genetics & development.* 2000;10(4):428-33. Epub 2000/07/13.

[24] Romanowski A, Migliori ML, Valverde C, Golombek DA. Circadian variation in Pseudomonas fluorescens (CHA0)-mediated paralysis of Caenorhabditis elegans. *Microbial pathogenesis.* 2011;50(1):23-30. Epub 2010/10/05.

[25] Migliori ML, Romanowski A, Simonetta SH, Valdez D, Guido M, Golombek DA. Daily variation in melatonin synthesis and AA-NAT activity in the nematode *Caenorhabditis elegans*. *Journal of pineal research.* 2011;*in press*.

[26] Tanaka D, Furusawa K, Kameyama K, Okamoto H, Doi M. Melatonin signaling regulates locomotion behavior and homeostatic states through distinct receptor pathways in Caenorhabditis elegans. *Neuropharmacology.* 2007;53(1):157-68. Epub 2007/06/19.

[27] Janssen T, Husson SJ, Meelkop E, Temmerman L, Lindemans M, Verstraelen K, et al. Discovery and characterization of a conserved pigment dispersing factor-like neuropeptide pathway in Caenorhabditis elegans. *Journal of neurochemistry*. 2009;111(1):228-41. Epub 2009/08/19.

[28] Meelkop E, Temmerman L, Schoofs L, Janssen T. Signalling through pigment dispersing hormone-like peptides in invertebrates. *Progress in neurobiology*. 2011;93(1):125-47. Epub 2010/11/03.

[29] Janssen T, Lindemans M, Meelkop E, Temmerman L, Schoofs L. Coevolution of neuropeptidergic signaling systems: from worm to man. *Annals of the New York Academy of Sciences*. 2010;1200:1-14. Epub 2010/07/17.

[30] Janssen T, Husson SJ, Lindemans M, Mertens I, Rademakers S, Ver Donck K, et al. Functional characterization of three G protein-coupled receptors for pigment dispersing factors in Caenorhabditis elegans. *The Journal of biological chemistry*. 2008;283(22):15241-9. Epub 2008/04/09.

[31] Simonetta SH, Golombek DA. An automated tracking system for Caenorhabditis elegans locomotor behavior and circadian studies application. *Journal of neuroscience methods*. 2007;161(2):273-80. Epub 2007/01/09.

[32] Buckingham SD, Sattelle DB. Strategies for automated analysis of C. elegans locomotion. *Invertebrate neuroscience: IN*. 2008;8(3):121-31. Epub 2008/08/09.

[33] Murphy JT, Bruinsma JJ, Schneider DL, Collier S, Guthrie J, Chinwalla A, et al. Histidine protects against zinc and nickel toxicity in Caenorhabditis elegans. *PLoS Genet*. 2011;7(3):e1002013. Epub 2011/04/02.

[34] Ye B, Rui Q, Wu Q, Wang D. Metallothioneins are required for formation of cross-adaptation response to neurobehavioral toxicity from lead and mercury exposure in nematodes. *PloS one*. 2010;5(11):e14052. Epub 2010/12/03.

[35] Wang D, Liu P, Yang Y, Shen L. Formation of a combined Ca/Cd toxicity on lifespan of nematode Caenorhabditis elegans. *Ecotoxicol Environ Saf*. 2010;73(6):1221-30. Epub 2010/06/29.

[36] Tvermoes BE, Boyd WA, Freedman JH. Molecular characterization of numr-1 and numr-2: genes that increase both resistance to metal-induced stress and lifespan in Caenorhabditis elegans. *J Cell Sci*. 2010;123(Pt 12):2124-34. Epub 2010/05/27.

[37] Schwartz MS, Benci JL, Selote DS, Sharma AK, Chen AG, Dang H, et al. Detoxification of multiple heavy metals by a half-molecule ABC

transporter, HMT-1, and coelomocytes of Caenorhabditis elegans. *PloS one.* 2010;5(3):e9564. Epub 2010/03/12.

[38] Au C, Benedetto A, Anderson J, Labrousse A, Erikson K, Ewbank JJ, et al. SMF-1, SMF-2 and SMF-3 DMT1 orthologues regulate and are regulated differentially by manganese levels in C. elegans. *PloS one.* 2009;4(11):e7792. Epub 2009/11/20.

[39] Zhang Y, Ye B, Wang D. Effects of metal exposure on associative learning behavior in nematode Caenorhabditis elegans. *Arch Environ Contam Toxicol.* 2010;59(1):129-36. Epub 2010/01/02.

[40] Chong R, Ke-zhou C, Zeng-liang Y. Induction of germline apoptosis by cobalt and relevant signal transduction pathways in Caenorhabditis elegans. *Toxicol Mech Methods.* 2009;19(9):541-6. Epub 2009/10/21.

[41] Rohlfing AK, Miteva Y, Hannenhalli S, Lamitina T. Genetic and physiological activation of osmosensitive gene expression mimics transcriptional signatures of pathogen infection in C. elegans. *PloS one.* 2010;5(2):e9010. Epub 2010/02/04.

[42] Choe KP, Strange K. Molecular and genetic characterization of osmosensing and signal transduction in the nematode Caenorhabditis elegans. *FEBS J.* 2007;274(22):5782-9. Epub 2007/10/20.

[43] Lamitina T, Huang CG, Strange K. Genome-wide RNAi screening identifies protein damage as a regulator of osmoprotective gene expression. *Proceedings of the National Academy of Sciences of the United States of America.* 2006;103(32):12173-8. Epub 2006/08/02.

[44] Lamitina ST, Morrison R, Moeckel GW, Strange K. Adaptation of the nematode Caenorhabditis elegans to extreme osmotic stress. *American journal of physiology Cell physiology.* 2004;286(4):C785-91. Epub 2003/12/03.

[45] Erkut C, Penkov S, Khesbak H, Vorkel D, Verbavatz JM, Fahmy K, et al. Trehalose Renders the Dauer Larva of Caenorhabditis elegans Resistant to Extreme Desiccation. *Current biology: CB.* 2011;21(15):1331-6. Epub 2011/07/26.

[46] Gal TZ, Glazer I, Koltai H. Stressed worms: responding to the post-genomics era. *Mol Biochem Parasitol.* 2005;143(1):1-5. Epub 2005/06/16.

[47] Van Raamsdonk JM, Hekimi S. Reactive Oxygen Species and Aging in Caenorhabditis elegans: Causal or Casual Relationship? *Antioxid Redox Signal.* 2010;13(12):1911-53. Epub 2010/06/24.

[48] Gems D, Doonan R. Antioxidant defense and aging in C. elegans: is the oxidative damage theory of aging wrong? *Cell Cycle.* 2009;8(11):1681-7. Epub 2009/05/05.

[49] Cabreiro F, Ackerman D, Doonan R, Araiz C, Back P, Papp D, et al. Increased life span from overexpression of superoxide dismutase in Caenorhabditis elegans is not caused by decreased oxidative damage. *Free Radic Biol Med.* 2011;51(8):1575-82. Epub 2011/08/16.

[50] Powell-Coffman JA. Hypoxia signaling and resistance in C. elegans. *Trends Endocrinol Metab.* 2010;21(7):435-40. Epub 2010/03/26.

[51] LaRue BL, Padilla PA. Environmental and genetic preconditioning for long-term anoxia responses requires AMPK in Caenorhabditis elegans. *PloS one.* 2011;6(2):e16790. Epub 2011/02/10.

[52] Hayakawa T, Kato K, Hayakawa R, Hisamoto N, Matsumoto K, Takeda K, et al. Regulation of anoxic death in Caenorhabditis elegans by mammalian apoptosis signal-regulating kinase (ASK) family proteins. *Genetics.* 2011;187(3):785-92. Epub 2011/01/08.

[53] Treinin M, Shliar J, Jiang H, Powell-Coffman JA, Bromberg Z, Horowitz M. HIF-1 is required for heat acclimation in the nematode Caenorhabditis elegans. *Physiol Genomics.* 2003;14(1):17-24. Epub 2003/04/11.

[54] Prahlad V, Cornelius T, Morimoto RI. Regulation of the cellular heat shock response in Caenorhabditis elegans by thermosensory neurons. *Science.* 2008;320(5877):811-4. Epub 2008/05/10.

[55] Edwards SL, Charlie NK, Milfort MC, Brown BS, Gravlin CN, Knecht JE, et al. A novel molecular solution for ultraviolet light detection in Caenorhabditis elegans. *PLoS biology.* 2008;6(8):e198. Epub 2008/08/09.

[56] Stergiou L, Eberhard R, Doukoumetzidis K, Hengartner MO. NER and HR pathways act sequentially to promote UV-C-induced germ cell apoptosis in Caenorhabditis elegans. *Cell Death Differ.* 2011;18(5):897-906. Epub 2010/12/15.

[57] Wang D, Liu P, Xing X. Pre-treatment with mild UV irradiation increases the resistance of nematode Caenorhabditis elegans to toxicity on locomotion behaviors from metal exposure. *Environ Toxicol Pharmacol.* 2010;29(3):213-22. Epub 2010/05/01.

[58] Murakami S, Johnson TE. A genetic pathway conferring life extension and resistance to UV stress in Caenorhabditis elegans. *Genetics.* 1996;143(3):1207-18. Epub 1996/07/01.

[59] Ewbank JJ. Signaling in the immune response. *WormBook: the online review of C elegans biology.* 2006:1-12. Epub 2007/12/01.

[60] Young JA, Dillin A. MAPping innate immunity. *Proceedings of the National Academy of Sciences of the United States of America.* 2004;101(35):12781-2. Epub 2004/08/26.

[61] Darby C, Cosma CL, Thomas JH, Manoil C. Lethal paralysis of Caenorhabditis elegans by Pseudomonas aeruginosa. *Proceedings of the National Academy of Sciences of the United States of America.* 1999;96(26):15202-7. Epub 1999/12/28.

[62] Gallagher LA, Manoil C. Pseudomonas aeruginosa PAO1 kills Caenorhabditis elegans by cyanide poisoning. *Journal of bacteriology.* 2001;183(21):6207-14. Epub 2001/10/10.

[63] Zhang L, Weng W, Guo J. Posttranscriptional mechanisms in controlling eukaryotic circadian rhythms. *FEBS letters.* 2011; 585(10):1400-5. Epub 2011/03/19.

[64] Harms E, Kivimae S, Young MW, Saez L. Posttranscriptional and posttranslational regulation of clock genes. *Journal of biological rhythms.* 2004;19(5):361-73. Epub 2004/11/10.

[65] Reischl S, Kramer A. Kinases and phosphatases in the mammalian circadian clock. *FEBS letters.* 2011;585(10):1393-9. Epub 2011/03/08.

In: Abiotic Stress: New Research ISBN: 978-1- 61942-197-4
Editors: Nikhil Haryana and Shreya Punj 

*Chapter 6*

# ABIOTIC STRESS RESPONSES IN PLANTS

***Yuriko Osakabe*[1] *and Keishi Osakabe*[2]**

[1] Gene Discovery Research Group, RIKEN Plant Science Center, 3-1-1 Kouyadai, Tsukuba, Ibaraki 305-0074, Japan

[2] Institute for Environmental Science and Technology, Saitama University, 225 Shimo-ohkubo, Sakura-ku, Saitama, Japan

## ABSTRACT

Plant growth and productivity are affected by environmental stresses, including drought, high salinity, and high and low temperature. Plants percept the stress signals and response to adapt these stresses. Plant abiotic stress responses are regulated by the integration of various environmental and endogenous signals including plant hormones. Plant adaptation to abiotic stresses is controlled by molecular cascades, in which expression of many genes are modulated, and also involved in the cross-talk pathway with other signaling mechanism. In this chapter, we show an overview of the current knowledge on plant molecular responses and adaptation to abiotic stresses.

## INTRODUCTION

Abiotic stresses, such as salinity, drought, temperature, and oxidative stress, cause major loss in crop yields. Plant engineering to develop stress tolerance becomes a necessity to increase plant biomass. Abiotic stress affects plant growth and productivity involves various biochemical and molecular events. Recent studies showed the complex cascades altering gene expression in response to abiotic stresses [1-6]. Stress affects plant growth and development, because plants need to re-allocate energy to be able to adapt to the stress [7, 8]. Engineering stress-tolerant plants requires a thorough understanding of various molecular responses to stress, especially of how the plant survives and continues to grow during the stress. This chapter will summarize the progress in research on abiotic stress responses in plants and focus on mechanisms changing gene expression involving transcription factors.

# 1. DROUGHT STRESS TRANSCRIPTION FACTORS

Drought stress affects various cellular and molecular events in plants, particularly expression of a variety of genes involved in not only stress tolerance but also stress signaling. Transcriptome analyses indicated that more than 300 transcripts were modulated by drought stress [9]. Drought and high-salinity stresses, are regulated by orchestrated signaling networks and found to cause similar changes in the transcriptome in *Arabidopsis* [2, 3]. ABA plays an important role in drought and salinity stress. ABA is dramatically accumulated in plant cells and activates the transcription of various genes under drought and salinity stress conditions. Drought and salinity stress signaling and ABA signaling are integrated into a complex regulatory network.

The transcriptional control that is operated under the stress signaling was controlled by the important transcription factors in drought stress response. The drought-responsive *cis*-element (DRE, 5'-TACCGACAT-3') has been identified in the promoter regions of drought and salinity stress-inducible genes. A part of DRE (5'-CCGAC-3') has also been reported to function as the *cis*-element that regulates transcription of various cold-responsive genes, and the *cis*-sequence was designated as C-repeat (CRT) and the low temperature-responsive element (LTRE) [10-13]. The transcription factors that can bind to the DRE/CRT sequence have been identified using yeast one-hybrid screening, as DREB1/CBF (DRE-binding protein 1/CRT-binding factor) and DREB2 [12, 14]. The DREB/CBF family is a subfamily of the plant-specific AP2 (APETALA2)/ERF transcription factor family [15]. The DREB1/CBF functions specifically in cold stress response in *Arabidopsis*. The expressions of *DREB2* genes, *DREB2A* and *DREB2B*, were induced by both dehydration and salinity stress but not cold stress and involved in drought- and salinity-stress responses [14-16]. The overexpression of *DREB2A* under the control of the cauliflower mosaic virus 35S RNA (35S) promoter did not affect the stress tolerance in the transgenic *Arabidopsis*. The further characterization of DREB2A protein showed that post-translational modification of the DREB2A is necessary for full activation of the protein [14]. The negative regulatory domain (NRD) exists in the central region of DREB2A and deletion of the NRD domain results in production of a constitutive active form of DREB2A [17]. The transgenic *Arabidopsis* plants overexpressing the constitutive active form showed a strong tolerance to drought stress with upregulation of the various stress-responsive genes [17]. The interacting protein with the DREB2A NRD has recently been identified and named DRIP1 (DREB2A interacting protein 1), which is a C3HC4 RING domain-containing protein [18]. Thus, the negative regulation of DREB2A activity is the proteolysis of DREB2A in 26S proteosome pathway mediated by DRIP1 [18]. The constitutive active form of DREB2A upregulated the genes include not only drought and salinity stress–responsive genes but also genes for heat shock proteins (HSPs) [19]. The transgenic *Arabidopsis* plants of the constitutive active form of DREB2A also showed an increased tolerance to heat stress [19]. The results suggested the presence of complex regulatory networks in cross-talk signaling pathway under the various stress.

Drought and salinity stresses are partly mediated by a phytohormone, abscisic acid (ABA) that induces expression of various genes through the ABA-responsive *cis*-element, ABRE. Both DRE/CRT and ABRE have been found in many stress-responsive gene promoters [3]. ABF/AREB (ABRE-binding factors /ABRE-binding proteins), bZIP-type transcription factors, have been identified from *Arabidopsis* [20, 21]. Overexpression of ABF3 or ABF4/AREB2 in *Arabidopsis* increased the expression levels of ABA-responsive genes, such as LEA genes, and resulted in the increased sensitivity to ABA and drought-stress tolerance [22].

The ABA-dependent phsopholylation was speculated to be involved in the activation of AREB/ABF proteins. The triple mutant, *snrk2.2snrk2.3snrk2.6,* the genes encoding type-2 SNF1-related protein kinases, showed severe phenotypes in the ABA sensitivity, phosphorylation of ABF/AREBs, and ABA-dependent gene expression [23-26]. Thus, SnRKs was involved in activation of AREB/ABFs. It has recently been shown that ABA promotes the interaction between ABA receptors, PYR/PYL/RCARs, and protein phsophatase 2Cs (PP2Cs), ABI1 and ABI2, which are negative regulators in ABA signaling [27, 28], and then inhibits PP2Cs activity to control ABA signaling [29, 30].

## 2. Cold Stress Transcription Factors

The expression of DREB1/CBFs (DREB1A-C/CBF1-3) is induced specifically by cold stress [14, 16], and they control the transcription of cold stress-responsive genes in *Arabidopsis*. The overexpression of DREB1B/CBF1 under the control of the 35S promoter triggered upregulation of various stress-responsive genes including LEA protein and increased tolerance to not only cold stress but also drought and salinity stresses [14, 31, 32]. Furthermore, overexpression of *DREB1/CBF* genes also increased tolerance to freezing, chilling, and drought stress in various plant species, including *Brassica napus*, tobacco, tomato, rice, wheat and maize [33-40].

*ICE1* (*inducer of CBF expression 1*), *HOS1* (*high expression of osmotically responsive gene 1*) and *MYB15* were found to encode factors involved in the regulatory mechanism for expression of DREB1/CBF in response to the cold stress. ICE1, a basic helix-loop-helix transcription factor, is a positive regulator of *CBF3*, and binds to the *cis*-regulatory elements in the *CBF3* promoter and activates the transcription [41, 42]. On the other hand, HOS1 is a RING-type ubiquitin E3 ligase that negatively regulates *DREB1/CBF* gene expression [43]. MYB15 that interacts with ICE1 is another negative regulator that binds the *CBF* promoters [44]. Furthermore, CAMTA3 (calmodulin binding transcription activator3) has been shown to be a positive regulator of *DREB1C/CBF2* gene [45]. The *camta3* mutants impaired the induction of *CBF2* in cold stress, and the *camta1camta3* double mutant showed a reduced freezing tolerance. This study suggested a link of calcium signaling and cold-regulated gene expression.

## 3. The Signal Transduction in Drought Stress Response

Recent studies revealed that AtHK1/AHK1 has roles in perception of drought stress. AtHK1/AHK1 is a histidine kinase in the His-Asp phosphorelay signaling system, which also mediates osmotic stress signaling in prokaryotes [46, 47]. AHK1 overexpression improves tolerance to drought stress in *Arabidopsis* [47]. A receptor-like kinase, RPK1, affects several ABA responses in *Arabidopsis*, was also found to be involved in the oxidative stress responses in plant cells [6]. The molecular mechanism for drought stress signaling, such as downstream pathways of AHK1 and RPK1, is currently mostly unknown.

Reactive oxygen species (ROS) production is induced by abiotic stresses and ROS detoxification is one of important steps in stress tolerance. ROS have also been implicated as important second messengers of ABA signaling in guard cells [48, 49, 50]. The RPK1 functions as a mediator the reactive oxygen species (ROS) homeostasis and also regulates abscisic acid (ABA) signaling in *Arabidopsis* [4, 6]. RPK1 overproduction in transgenic plants increased their tolerance to drought and oxidative stresses. Therefore, these findings provide the evidence of the cross-talk between water stress and other stress signaling mechanisms, implying that stress response in plants is under the control of a complex regulatory network.

## REFERENCES

[1] Wang W, Vinocur B, Altman A (2003) Plant responses to drought, salinity and extreme temperatures: towards genetic engineering for stress tolerance. *Planta,* 218, 1-14.

[2] Vinocur, B. and Altman, A. (2005) Recent advances in engineering plant tolerance to abiotic stress: achievements and limitations. *Curr. Opin. Biotechnol.*, 16, 123-132.

[3] Yamaguchi-Shinozaki, K. and Shinozaki, K. (2006) Transcriptional regulatory networks in cellular responses and tolerance to dehydration and cold stresses. *Annu. Rev. Plant Biol.*, 57, 781–803.

[4] Osakabe Y, Maruyama K, Seki M, Satou M, Shinozaki K, Yamaguchi-Shinozaki K (2005) Leucine-rich repeat receptor-like kinase1 is a key membrane-bound regulator of abscisic acid early signaling in *Arabidopsis*. *Plant Cell,* 17, 1105-1119.

[5] Mittler R, Vanderauwera S, Gollery M, Van Breusegem F (2004) Reactive oxygen gene network of plants. *Trends Plant Sci.*, 9, 490-498.

[6] Osakabe Y, Mizuno S, Tanaka H, Maruyama K, Osakabe K, Todaka D, Fujita Y, Kobayashi M, Shinozaki K, Yamaguchi-Shinozaki K (2010) Overproduction of the membrane-bound receptor-like protein kinase 1, RPK1, enhances abiotic stress tolerance in *Arabidopsis*. *J. Biol. Chem.*, 285, 9190-9201.

[7] Ahuja I, de Vos RC, Bones AM, Hall RD. (2010) Plant molecular stress responses face climate change. *Trends Plant Sci.*, 15, 664-674.

[8] Skirycz A, Inze D (2010) More from less: plant growth under limited water. *Curr. Opinion Biotechnol.,* 21, 197-203.

[9] Seki, M., Narusaka, M., Ishida, J., Nanjo, T., Fujita, M., Oono, Y., Kamiya, A., Nakajima, M., Enju, A., Sakurai, T., Satou, M., Akiyama, K., Taji, T.,

Yamaguchi-Shinozaki, K., Carninci, P., Wakasa, J., Hayashizaki, H., Shinozaki, K. (2002) Monitoring the expression profiles of 7000 *Arabidopsis* genes under drought, cold and high-salinity stresses using a full-length cDNA microarray. *Plant J.*, 31, 279–292.

[10] Baker, S.S., Wilhelm, K.S. Thomashow, M.F. (1994) The 5'-region of *Arabidopsis thaliana cor15a* has *cis*-acting elements that confer cold-, drought-, and ABA-regulated gene expression. *Plant Mol. Biol.*, 24, 701-713.

[11] Jiang, C., Iu, B., Singh, J. (1996) Requirement of a CCGAC *cis*-acting element for cold induction of the *BN115* gene from winter *Brassica napus*. *Plant Mol. Biol.*, 30, 679–684.

[12] Stockinger,E.J., Gilmour, S.J., Thomashow, M.F. (1997) *Arabidopsis thaliana CBF1* encodes an AP2 domain-containing transcription activator that binds to the C-repeat/DRE, a *cis*-acting DNA regulatory element that stimulates transcription in response to low temperature and water deficit. *Proc. Natl. Acad. Sci. USA*, 94, 1035-1040.

[13] Thomashow, M.F. (1999) Plant cold acclimation: freezing tolerance genes and regulatory mechanisms. *Annu. Rev. Plant Physiol. Plant Mol. Biol.*, 50, 571−599.

[14] Liu, Q., Kasuga, M., Sakuma, Y., Abe, H., Miura, S., Yamaguchi-Shinozaki, K., Shinozaki, K. (1998) Two transcription factors, DREB1 and DREB2, with an EREBP/AP2 DNA binding domain separate two cellular signal transduction pathways in drought- and low-temperature responsive gene expression, respectively, in Arabidopsis. *Plant Cell*, 10, 1391–1406.

[15] Sakuma, Y., Liu, Q., Dubouzet, J.G., Abe, H., Shinozaki, K., Yamaguchi-Shinozaki, K. (2002) DNA-binding specificity of the ERF/AP2 domain of *Arabidopsis* DREBs, transcription factors involved in dehydration- and cold-inducible gene expression. *Biochem. Biophys. Res. Commun.*, 290, 998–1009.

[16] Nakashima, K., Shinwari, Z.K., Sakuma, Y., Seki, M., Miura, S., Shinozaki, K., Yamaguchi-Shinozaki, K. (2000) Organization and expression of two *Arabidopsis DREB2* genes encoding DRE-binding proteins involved in dehydration- and high-salinity-responsive gene expression. *Plant Mol. Biol.*, 42, 657–665.

[17] Sakuma, Y., Maruyama, K., Osakabe, Y., Qin, F., Seki, M., Shinozaki, K., Yamaguchi-Shinozaki, K. (2006) Functional analysis of an *Arabidopsis* transcription factor, DREB2A, involved in drought-responsive gene expression. *Plant Cell*, 18, 1292–1309.

[18] Qin, F., Sakuma, Y., Tran, L.-S.P., Maruyama, K., Kidokoro, S., Fujita, Y., Fujita, M., Umezawa, T., Sawano, Y., Miyazono, K., Tanokura, M., Shinozaki, K., Yamaguchi-Shinozaki, K. (2008) *Arabidopsis* DREB2A-interacting proteins function as RING E3 ligases and negatively regulate plant drought stress-responsive gene expression. *Plant Cell*, 20, 1693–1707.

[19] Sakuma, Y., Maruyama, K., Qin, F., Osakabe, Y., Shinozaki, K., Yamaguchi-Shinozaki, K.. (2006) Dual function of an *Arabidopsis* transcription factor DREB2A in water-stress-responsive and heat-stress-responsive gene expression. *Proc. Natl. Acad. Sci. USA*, 103, 18822–18827.

[20] Choi, H.-i., Hong, J.-h., Ha, J.-o., Kang, J.-y., Kim, S.Y., (2000) ABFs, a family of ABA-responsive element binding factors. *J. Biol. Chem.*, 275, 1723-1730.

[21] Uno, Y., Furihata, T., Abe, H., Yoshida, R., Shinozaki, K., Yamaguchi-Shinozaki, K. (2000) *Arabidopsis* basic leucine zipper transcription factors involved in an abscisic acid-dependent signal transduction pathway under drought and high-salinity conditions. *Proc. Natl. Acad. Sci. USA*, 97, 11632–11637.

[22] Kang, J.-y., Choi, H.-i., Im, M.-y., Kim, S.Y. (2002) Arabidopsis basic leucine zipper proteins that mediate stress-responsive abscisic acid signaling. *Plant Cell*, 14, 343–357.

[23] Fujii, H., Zhu, J.-K. (2009) Arabidopsis mutant deficient in 3 abscisic acid-activated protein kinases reveals critical roles in growth, reproduction, and stress. *Proc. Natl. Acad. Sci. USA*, 106, 8380-8385.

[24] Fujita, Y., Nakashima, K., Yoshida, T., Katagiri, T., Kidokoro, S., Kanamori, N., Umezawa, T., Fujita, M., Maruyama, K., Ishiyama, K., Kobayashi, M., Nakasone, S., Yamada, K., Ito, T., Shinozaki, K., Yamaguchi-Shinozaki, K. (2009) Three SnRK2 protein kinases are the main positive regulators of abscisic acid signaling in response to water stress in Arabidopsis. *Plant Cell Physiol.*, 50, 2123-2132.

[25] Nakashima, K., Fujita, Y., Kanamori, N., Katagiri, T., Umezawa, T., Kidokoro, S., Maruyama, K., Yoshida, T., Ishiyama, K., Kobayashi, M., Shinozaki, K., Yamaguchi-Shinozaki, K. (2009) Three Arabidopsis SnRK2 protein kinases, SRK2D/SnRK2.2, SRK2E/SnRK2.6/OST1 and SRK2I/SnRK2.3, involved in ABA signaling are essential for the control of seed development and dormancy. *Plant Cell Physiol.*, 50, 1345-1363.

[26] Umezawa, T., Sugiyama, N., Mizoguchi, M., Hayashi, S., Myouga, F., Yamaguchi-Shinozaki, K., Ishihama, Y., Hirayama, T., Shinozaki, K. (2009) Type 2C protein phosphatases directly regulate abscisic acid-

activated protein kinases in Arabidopsis. *Proc. Natl. Acad. Sci. USA*, 10, 17588-17593.

[27] Finkelstein, R.R., Gampala, S.S.L., Rock, C.D. (2002) Abscisic acid signaling in seeds and seedlings. *Plant Cell*, 13, S15-S45.

[28] Sheen J. (1998) Mutational analysis of protein phosphatase 2C involved in abscisic acid signal transduction in higher plants. *Proc. Natl. Acad. Sci. USA*, 95, 975-980.

[29] Ma, Y., Szostkiewicz, I., Korte, A., Moes, D., Yang, Y., Christmann, A., Grill, E. (2009) Regulators of PP2C phosphatase activity function as abscisic acid sensors. *Science*, 324, 1064-1068.

[30] Park, S.-Y., Fung, P., Nishimura, N., Jensen, D.R., Fujii, H., Zhao, Y., Lumba, S., Santiago, J., Rodrigues, A., Chow, T.-f.F., Alfred, S.E., Bonetta, D., Finkelstein, R., Provart, N.J., Desveaux, D., Rodriguez, P.L., McCourt, P., Zhu, J.-K., Schroeder, J.I., Volkman, B.F., Cutler, S.R. (2009) Abscisic acid inhibits type 2C protein phosphatases via the PYR/PYL family of START proteins. *Science*, 324, 1068-1071.

[31] Kasuga, M., Liu, Q., Miura, S., Yamaguchi-Shinozaki, K., Shinozaki, K. (1999) Improving plant drought, salt, and freezing tolerance by gene transfer of a single stress-inducible transcription factor. *Nat. Biotechnol.*, 17, 287–291.

[32] Gilmour, S.J., Sebolt, A.M., Salazar, M.P., Everard, J.D., Thomashow, M.F. (2000) Overexpression of the Arabidopsis *CBF3* transcriptional activator mimics multiple biochemical changes associated with cold acclimation. *Plant Physiol.*, 124, 1854–1865.

[33] Jaglo, K.R., Kleff, S., Amundsen, K.L., Zhang, X., Haake, V., Zhang, J.Z., Deits, T., Thomashow, M.F. (2001) Components of the Arabidopsis C-repeat/dehydration-responsive element binding factor cold-response pathway are conserved in *Brassica napus* and other plant species. *Plant Physiol.*, 127, 910–917.

[34] Kasuga, M., Miura, S., Shinozaki, K., Yamaguchi-Shinozaki, K. (2004) A combination of the *Arabidopsis* DREB1A gene and stress-inducible *rd29A* promoter improved drought- and low-temperature stress tolerance in tobacco by gene transfer. *Plant Cell Physiol.*, 45, 346-350.

[35] Hsieh, T.-H., Lee, J.-T., Charng, Y.-Y., Chan, M.-T. (2002) Tomato plants ectopically expressing Arabidopsis CBF1 show enhanced tolerant to water-deficit stress. *Plant Physiol.*, 130, 618-626.

[36] Ito, Y., Katsura, K., Maruyama, K., Taji, T., Kobayashi, M., Seki, M., Shinozaki, K., Yamaguchi-Shinozaki, K. (2006) Functional analysis of rice

DREB1/CBF-type transcription factors involved in cold-responsive gene expression in transgenic rice. *Plant Cell Physiol.*, 47, 141–153.

[37] Oh, S.-J., Song, S.I., Kim, Y.S., Jang, H.-J., Kim, S.Y., Kim, M., Kim, Y.-K., Nahm, B.H., Kim, J.-K. (2005) Arabidopsis CBF3/DREB1A and ABF3 in transgenic rice increased tolerance to abiotic stress without stunting growth. *Plant Physiol.*, 138, 341–351.

[38] Pellegrineschi, A., Reynolds, M., Pacheco, M., Brito, R.M., Almeraya, R., Yamaguchi-Shinozaki, K., Hoisington, D. (2004) Stress-induced expression in wheat of the *Arabidopsis thaliana* DREB1A gene delays water stress symptoms under greenhouse conditions. *Genome* 47, 493–500.

[39] Qin, F., Sakuma, Y., Li, J., Liu, Q., Li, Y.-Q., Shinozaki, K., Yamaguchi-Shinozaki, K. (2004) Cloning and functional analysis of a novel DREB1/CBF transcription factor involved in cold-responsive gene expression in *Zea mays* L. *Plant Cell Physiol.*, 45, 1042–1052.

[40] Zhang, J.Z., Creelman, R.A., Zhu, J.-K. (2004). From laboratory to field. Using information from Arabidopsis to engineer salt, cold, and drought tolerance in crops. *Plant Physiol.*, 135, 615-621.

[41] Chinnusamy, V., Ohta, M., Kanrar, S., Lee, B.-h., Hong, X., Agarwal, M., Zhu, J.-K. (2003) ICE1: a regulator of cold-induced transcriptome and freezing tolerance in *Arabidopsis*. *Genes Dev.*, 17, 1043-1054.

[42] Lee, B.-h., Henderson, D.A., Zhu, J.-K. (2005) The *Arabidopsis* cold-responsive transcriptome and its regulation by ICE1. *Plant Cell*, 17, 3155-3175.

[43] Ishitani, M., Xiong, L., Lee, H., Stevenson, B., Zhu, J.-K. (1998) *HOS1*, a genetic locus involved in cold-responsive gene expression in Arabidopsis. *Plant Cell*, 10, 1151-1162.

[44] Agarwal, M., Hao, Y., Kapoor, A., Dong, C.-H., Fujii, H., Zheng, X. (2006) A R2R3 type MYB transcription factor is involved in the cold regulation of CBF genes and in acquired freezing tolerance. *J. Biol, Chem.,* 281, 37636-37645.

[45] Doherty, C.J., Van Buskirk, H.A., Myers, S.J., Thomashow, M.F. (2009) Roles for *Arabidopsis* CAMTA transcription factors in cold-regulated gene expression and freezing tolerance. *Plant Cell*, 21, 972–984.

[46] Urao, T., Yakubov, B., Satoh, R., Yamaguchi-Shinozaki, K., Seki, M., Hirayama, T., Shinozaki, K. (1999) A transmembrane hybrid-type histidine kinase in Arabidopsis functions as an osmosensor. *Plant Cell*, 11, 1743-1754.

[47] Tran, L.-S.P., Urao, T., Qin, F., Maruyama, K., Kakimoto, T., Shinozaki, K., Yamaguchi-Shinozaki, K. (2007) Functional analysis of AHK1/ATHK1

and cytokinin receptor histidine kinases in response to abscisic acid, drought, and salt stress in *Arabidopsis*. *Proc. Natl. Acad. Sci. USA.*, 104, 20623-20628.

[48] Kwak, J.M., Mori, I., Pei, Z.-M., Leonhardt, N., Torres, M.A., Dangl, J.L., Bloom, R., Bodde, S., Jones, J.D.G., and Schroeder, J.I. (2003) NADPH oxidase AtrbohD and AtrbohF genes function in ROS-dependent ABA signaling in Arabidopsis. *EMBO J.,* 22, 2623-2633.

[49] Murata, Y., Pei, Z.-M., Mori, I.C., and Schroeder, J.I. (2001) Arabidopsis OST1 Protein Kinase Mediates the Regulation of Stomatal Aperture by Abscisic Acid and Acts Upstream of Reactive Oxygen Species Production. *Plant Cell,* 13, 2513-2523.

[50] Neill, S., Desikan, R., and Hancock, J. (2002) Hydrogen peroxide signaling. *Curr. Opin. Plant Biol.* 5, 388-395.

In: Abiotic Stress: New Research ISBN: 978-1- 61942-197-4
Editors: Nikhil Haryana and Shreya Punj 

*Chapter 7*

# PLANT CORTICAL MICROTUBULES AND ABIOTIC STRESS SIGNALS

***Che Wang****

[1] Biological Science and Technology College, Shenyang Agricultural University, Shenyang, PR China

## ABSTRACT

High soil salinity, drought, and low temperature response mechanisms have been studied intensively. Recently, significant progress has been made in elucidating the nature of microtubules (MTs) that are believed to be intimately involved in stress response mechanisms. In recent years, the known functions of MTs have expanded from being just static scaffolds to playing critical roles in almost all intracellular activities. Accordingly, research on MT arrays in the context of abiotic stresses may reveal important mechanisms of plant stress tolerance, and consequently these arrays have become a hot and key research topic. In this chapter, we review current knowledge on the relationship between plant cortical microtubules (CMTs) and abiotic stress signalling. The data show that at least three early response mechanisms, namely, abscisic acid (ABA) biosynthesis, calcium ion ($Ca2+$) signalling, and the formation of plant cell walls, follow CMT reorganization, and are mediated by dynamic changes in the CMTs.

---

* Corresponding author, Che Wang , Biological Science and Technology College, Shenyang Agricultural University, Shenyang 110161, PR China, Phone: (8624) 88487163, Fax: (8624) 88411127, E-mail: wangwangche@163.com

**Keywords:** Microtubules(MTs), Abiotic stresses, Signalling pathway

Microtubules (MTs), highly dynamic structures, are the main integrants of the cytoskeleton. They are assembled from heterodimers of α- and β-tubulin subunits to form four unique types of array—cortical MTs (CMTs), pre-prophase band (PPB), spindle, and phragmoplast—throughout the plant cell cycle [6, 31]. From the time when they were first discovered in 1963, the important functions of MT arrays have been identified in almost every intracellular activity, from cell division to cell movement, cell morphogenesis, and even cell signal transduction [4, 9]. Recently, research on MT arrays has expanded to encompass the responses to abiotic stresses, primarily those of high salt, drought, and low temperature [1, 12, 18, 22], constituting a new topic in the subject of abiotic stress tolerance mechanisms in plants. The effects of high salinity, drought, and low temperature on plants have traditionally received considerable attention because they are major stress factors that adversely affect crop productivity and quality [29, 33]. Furthermore, the organization and dynamics of MTs are likely to play a vital role in enhancing plant tolerance through various cell activities.

A generic stress signal transduction pathway starts with signal perception followed by the generation of second messengers and activity of cell-adaptive responses to abiotic stresses [29, 33]. It is of fundamental importance to understand the mechanisms by which plants perceive environmental stimuli and transmit signals to the cellular machinery. Although there have been significant advances in research on the adaptive mechanisms associated with the responses to high salinity, drought, and low temperature, there are still many facets of the signal transduction pathways that await elucidation [29, 33].

From the perspective of a single cell, an environmental signal is a packet of information transmitted from cell wall to plasma membrane to protoplasm. Because the CMTs underlying the plasma membrane in plant cells constitute a continuum with the cell wall and plasma membrane, it has been assumed for over 10 years that these CMTs may be responsible for perceiving environmental signals [6, 11, 16]. Recent advances have led to the discovery that CMTs rearrangements mediate cell responses to certain environmental signals; for example, gravity, wounding, light, and biotic stresses [11]. Thus, changes in CMT arrays are often associated with the regulatory mechanisms of abiotic stress signalling pathways. To date, CMTs have been shown to be involved in at least three signalling pathways, namely, ABA biosynthesis, Ca2+ signalling, and the formation of plant cell walls.

ABA is a major phytohormone that can enhance plant tolerance under abiotic stresses. Rapid biosynthesis and significant accumulation of ABA in plant cells is an early and more obvious physiological event in the adaptive responses to abiotic stresses [29, 33]. It has also been shown that ABA can affect the organization of MTs under normal conditions [19, 21]. Under abiotic stress (chilling temperature), a clear destruction of MTs and that anti-MT drug treatment significantly accelerates and enhances the chilling damage in cotton, whereas pretreatment with ABA can prevent chilling damage and MT destruction. Evidence for an ABA-induced decrease in the generation of MT deploymerization at the initial stages of plant cold acclimation has also been presented [17]. These findings suggest that ABA enhances plant cold tolerance partly through affecting MT organization. Most interestingly, MTs that interact with integrin-like proteins have been shown to be involved in cell wall–plasma membrane interaction and osmotic stress-induced ABA accumulation [15]. Pharmacological approaches provide further evidence that disruption and stabilization of MTs significantly stimulate ABA accumulation in maize root cells, suggesting that MT dynamics may be involved in mediating osmotic stress-induced ABA biosynthesis [14].

The Ca2+ signal generated by transient Ca2+ influx into the cell cytoplasm can be induced under salinity, drought, and cold stresses [29, 33]. Therefore, the channels responsible for this Ca2+ influx may represent one type of sensor for these stresses [29]. The results of several studies have shown that CMTs may be involved in regulating the activity of Ca2+ channels under normal conditions [25, 26]. Some studies have found that under abiotic stress, disruption of the CMT network is an important upstream event of Ca2+ signalling. For example, the disruption of the MT network by various MT-interacting drugs increases cytosolic Ca2+ in cold-shocked N. plumbaginifolia protoplasts [16]. Sangwan and colleagues report that cold activation of BN115 requires membrane rigidification, cytoskeleton reorganization, and Ca2+ influx. They proposed that in response to the changes, the cytoskeleton acts as a scaffold that transduces physical forces into biochemical signals. It is likely that cytoskeleton rearrangements may mediate the transduction of the cold signal from the rigidified membrane to the Ca2+ channels [20]. In addition, disruption of MTs can ensure the efficient functioning of cold-activated Ca2+ channels, thereby triggering the signal pathway, and culminating in cold acclimation [1]. By using yellow cameleon (YC.3.6) as a calcium reporter, notable increases in [Ca2+]cyt after initial salt stress-induced depolymerization of the CMTs is found, and these increases are inhibited when paclitaxel is added [27]. These results suggest that it is likely that a Ca2+ influx follows the changes in MT array responsible for regulating Ca2+ channels under abiotic stresses.

The cell wall is the first structure of a plant cell to perceive environmental stimuli; accordingly, the cell wall must contain elements responsible for the response mechanisms. During primary wall formation, CMTs establish transverse organization during the early stages of cell elongation to direct the orientation of cellulose microfibril deposition, aligning the CESA complexes [2, 3, 24]. However, some recent studies have established that the alignment of CMTs and cellulose microfibrils is not always coupled during cell elongation [23, 28]. This suggests that cellulose microfibrils have the ability to largely self-align in the absence of the pre-stage orientation of CMTs. Therefore, the CMTs may only be required for directing cellulose into an orientation, for example, in response to various environmental stimuli. Recent research on plant drought tolerance has demonstrated that CMT organization can drive cellulose microfibril arrays by guiding the trajectories of active CESA in the plasma membrane, and that the association of CESA compartments with CMTs is greatly increased during osmotic stress [13]. Another report has confirmed that CMTs not only guide the trajectories of CESA complexes (CSCs) but also that osmotic stress can affect internalization of CSCs in MT-associated CESA compartments [5]. Therefore, CMTs may constitute a target in the plant cell wall—a complex sensory panel for transmitting extracellular information to the intracellular signalling network—under abiotic stresses.

In addition, reactive oxygen species (ROS), as signal molecules induced by abiotic stresses in plants, have been shown to cause fragmentation of MTs in vitro and mitotic arrest in tobacco BY-2 cells [8,30]. Phospholipase D (PLD), which is involved in ABA- and ROS-mediated processes as part of the plant's response to abiotic stresses, has been considered a potential linker between CMTs and the plasma membrane [7, 10, 32]. The close relationship between these signal molecules and CMTs further suggests that CMTs may be involved in the responses to signal transduction pathways under abiotic stresses.

In summary, it is very clear that remodeling of CMTs, which constitute a continuum between cell wall and plasma membrane as the headstream of signal transduction, is an integral element in the signalling cascades that significantly enhance plant adaptation to abiotic stress. Although the mechanisms by which CMTs regulate abiotic stress signalling pathways and the identity of the MAPs involved in these mechanisms remain to be elucidated, we believe that accumulating evidence over the next few years will provide further support for the close relationship between CMTs and the process of abiotic stress signal.

# REFERENCES


[1] Abdrakhamanova A, Wang QY, Khokhlova L, Nick P (2003) Is microtubule disassembly a trigger for cold acclimation? Plant Cell Physiol 44: 676– 686

[2] Baskin TI (2001) On the alignment of cellulose microfibrils by cortical microtubules: a review and a model. Protoplasma 215: 150– 171

[3] Burk DH, Ye ZH (2002) Alteration of oriented deposition of cellulose microfibrils by mutation of a katanin-like microtubule-severing protein. Plant Cell 14 (9): 2145– 2160

[4] Buschmann H, Lloyd CW (2008) *Arabidopsis* Mutants and the Network of Microtubule- Associated Functions. Mol Plant 6: 888– 898

[5] Crowell EF, Bischoff V, Desprez T, Rolland A, Stierhof YDi, Schumacher K, Gonneau M, Haöfte H, Vernhettesa S (2009) Pausing of golgi bodies on microtubules regulates secretion of cellulose synthase complexes in *Arabidopsis*. Plant Cell 21: 1141– 1154

[6] Cyr RJ, Palevitz BA (1995) Organization of cortical microtubules in plant cells. Curr Opin Cell Biol 7: 65– 71

[7] Dhonukshe P, Laxalt AM, Goedhart J, Gadella TWJ, Munnik T (2003) Phospholipase D activation correlates with microtubule reorganization in living plant cells. Plant Cell 15: 2666– 2679

[8] Dixit R, Cyr R (2003) Cell damage and reactive oxygen species production induced by fluorescence microscopy: effect on mitosis and guidelines for non-invasive fluorescence microscopy. Plant J 36: 280– 290

[9] Fu Y, Gu Y, Zheng Z, Wasteneys G, Yang Z (2005) *Arabidopsis* interdigitating cell growth requires two antagonistic pathways with opposing action on cell morphogenesis. Cell, 120: 687– 700

[10] Gardiner JC, Harper JDI, Weerakoon ND, Collings DA, Ritchie S, Gilroy S, Cyr RJ, Marc J (2001) A 90-kD phospholipase D from tobacco binds to microtubules and the plasma membrane. Plant cell 13: 2143– 2158

[11] Gundersen GG, Cook TA (1999) Microtubules and signal transduction. Curr Opin Cell Biol 11: 81– 94

[12] Guo LB, Ho CM, Kong ZS, Lee YR, Qian Q, Bo L (2009) Evaluating the microtubule cytoskeleton and its interacting proteins in monocots by mining the rice genome. Ann Bot 103(3):387– 402

[13] Gutierrez R, Lindeboom JJ, Paredez AR, Emons AM, Ehrhardt DW (2009) *Arabidopsis* cortical microtubules position cellulose synthase delivery to the plasma membrane and interact with cellulose synthase trafficking compartments. Nat Cell Biol 11(7):797– 806

[14] Lü B, Gong ZH, Wang J, Zhang JH, Liang JS (2007b) Microtubule dynamics in relation to osmotic stress-induced ABA accumulation in Zea mays roots. J Exp Bot 58(10): 2565– 2572

[15] Lü B, Chen F, Gong ZH, Xie H, Liang J (2007a) Integrin-like protein is involved in the osmotic stress-induced abscisic acid biosynthesis in *Arabidopsis thaliana*. J In Plant Biol 49: 540–549

[16] Mazars C, Thion L, Thuleau P, Graziana A, Knight MR, Moreau M, Ranjeva R (1997) Organization of cytoskeleton controls the changes in cytosolic calcium of cold-shocked *Nicotiana plumbaginifolia* protoplasts. Cell Calcium 22:413—420

[17] Olinevich OV, Khokhlova LP, Raudaskoski M (2002) The microtubule stability increases in abscisic acid-treated and cold acclimated differentiating vascular root tissues of wheat. J plant physiol 159: 465– 472

[18] Ruggenthaler P, Fichtenbauer D, Krasensky J, Jonak C, Waigmann E (2009) Microtubule-Associated Protein AtMPB2C plays a role in organization of cortical microtubules, stomata patterning, and tobamovirus infectivity. Plant Physiol 149:1354– 1365

[19] Sakiyama M, Shibaoka H (1990) Effect of abscisic acid on the orientation and cold stability of cortical microtubules in epicotyl cells of the dwarf pea. Protoplasma 157: 165– 171

[20] Sangwan V, Foulds I, Singh J, Dhindsa RS (2001) Cold-activation of *Brassica napus* BN115 promoter is mediated by structural changes in membranes and cytoskeleton, and requires $Ca^{2+}$ influx. Plant J 27(1): 1– 12

[21] Shibaoka H (1994) Plant hormone-induced changes in the orientation of cortical microtubules. Annu Rev Plant Physiol Mol Biol 45: 527– 544

[22] Shoji T, Suzuki K, Abe T, Kaneko Y, Shi H, Zhu JK, Rus A, Hasegawa PM, Hashimoto T (2006) Salt stress affects cortical microtubule organization and helical growth in *Arabidopsis*. Plant Cell Physiol 47: 1158– 1168

[23] Sugimoto K, Himmelspach R, Williamson RE, Wasteneys GO (2003) Mutation or drug-dependent microtubule disruption causes radial swelling without altering parallel cellulose microfibril deposition in *Arabidopsis* root cells. Plant Cell 15: 1414– 1429

[24] Sugimoto K, Williamson RE, Wasteneys GO (2000) New techniques enable comparative analysis of microtubule orientation, wall texture, and growth rate in intact roots of *Arabidopsis*. Plant Physiol 124:1493–1506

[25] Thion L, Mazars C, Nacry P, Bouchez D, Moreau M, Ranjeva R, Thuleau P (1998) Plasma membrane depolarization-activated calcium channels, stimulated by microtubule depolymerizing drugs in wild-type *Arabidopsis*

*thaliana* protoplasts, display constitutively large activities and a longer half-life in *ton* 2 mutant cells affected in the organization of cortical microtubules. Plant J 13: 603–610

[26] Thion L, Mazars C, Thuleau P, Graziana A, Rossignol M, Moreau M, Ranjeva R (1996) Activation of plasma membrane voltage-dependent calcium-permeable channels by disruption of microtubules in carrot cells. FEBS Lett 393: 13–18

[27] Wang C, Li J, Yuan M (2007a) Salt tolerance requires cortical microtubule reorganization in *Arabidopsis*. Plant Cell Physiol 48 (11): 1534–1547

[28] Whittington AT, Vugrek O, Wei KJ, Hasenbein NG, Sugimoto K, Rashbrooke MC, Wasteneys GO (2001) MOR1 is essential for organizing cortical microtubules in plants. Nature 411: 610–613

[29] Xiong LM, Schumaker KS, Zhu JK (2002) Cell signaling during cold, drought, and salt Stress. Plant Cell 14: 165–183

[30] Xu C, Liu C, Guo H, Li Z, Jiang Y, Zhang D, Yuan M (2006) Photosensitive breakage of fluorescence-labeled microtubules and its mechanism. Acta Physica Sinica 55: 206–210 (in Chinese)

[31] Yuan M, Shaw PJ, Warn RM, Lloyd CW (1994) Dynamic reorientation of cortical microtubules, from transverse to longitudinal, in living plant cells. Proc Natl Acad Sci USA 91: 6050–6053

[32] Zhang WH, Yu LJ, Zhang YY, Wang XM (2005) Phospholipase D in the signaling networks of plant response to abscisic acid and reactive oxygen species. Biochim Biophys Acta 1736: 1–9

[33] Zhu JK (2002) Salt and drought stress signal transduction in plants. Annu. Rev. Plant Biol 53: 247–273

# INDEX

## B



## C

## D

## E



## F

## G

## H



## I



## J

## K



## L



## M

## P

## Q



## R

## S

## T